...pe you enjoy this book.

...rn or renew it by the due da...

...y it ... ww.norf... ...ov

...ary app ... ...r

RAILWAY BLUNDERS

Adrian Vaughan

Ian Allan

PUBLISHING

Bibliography

Newspapers
North West Evening Mail
The Independent
The Observer
The Times

Magazines
Modern Railways, 1965-2003
Rail, 1999-2003
Railwatch, 1998-2000
Railway Magazine, 1998/9
Missing Link (Wealden Line Campiagn)
Oxford & Bucks Action Committee newsletter

Hansard
Vol 211 (14.7.92), Cols 971-986
Vol 211 (29.10.92), Cols 1163-1220
Vol 216 (12.1.93), Cols 783-785
The Railways Act 1921, Cap 55 11 & 12 Geo V

National Archive
MT62/138

Proceedings of the Institute of Civil Engineers
Structural Accidents and their Causes. P. G. Silby BSc,
A. C. Walker PhD, 1977. Part 1 (May) pp191-208
Obit Thomas Bouch, Vol 63, 1881

Official Reports
Collapse of the NATM tunnels at Heathrow (HSE Books, 2000)

Books
Blueprints for Bankruptcy — Ted Gibbins
(Leisure Products, 1995)
British Locomotives of the 20th Century — O. S. Nock
(Patrick Stephens Ltd, 1984)
British Rail after Beeching — G. Freeman Allen
(Ian Allan Publishing, 1966)
British Rail Main Line Diesel Locomotives (revised edition) —
Colin J. Marsden and Graham B. Fenn
(Oxford Publishing Co, 2000)
British Railways Engineering — John Johnson and Robert A. Long,
edited by Roland C. Bond (MEP Publications, 1981)
British Railway History 1830-1876 and *British Railway History
1877-1947* — C. Hamilton Ellis (George Allen & Unwin, 1956)
60 Years in Steam — D. W. Harvey (David & Charles, 1986)
Leader — Kevin Robertson (Alan Sutton, 1988)
The Midland Railway — C. Hamilton Ellis
(Ian Allan Publishing, 1955)
Railwaymen, Politics and Money — Adrian Vaughan
(John Murray, 1997)
Richard Trevithick, Giant of Steam — Anthony Burton
(Aurum Press, 2000)
The Tay Bridge Disaster — John Thomas
(David & Charles, 1972)
Timothy Hackworth and the Locomotive — Robert Young MIME
(Shildon Jubilee Committee, 1975)
Tracks to Disaster — Adrian Vaughan (Ian Allan Publishing, 2000)
Transport in Britain — Philip Bagwell & Peter Lyth (Hamilton &
London, 2002)

Previous page: The 'Blue Pullman' sets, introduced on the Western and London Midland Regions, were designed to provide high-quality services and were in many respects the forerunners of the highly successful HSTs developed in the 1970s. In terms of ride, however, the units were poor and were destined to have relatively short lives. This view, taken in December 1966, shows one of the Western Region units in service at the west end of Foxes Wood Tunnel.
Adrian Vaughan Collection

Front cover top: The Tay Bridge from the north after the High Girders fell. *British Railways*

Front cover centre: A steam crane between duties. *Ian Allan Library*

Front cover bottom: A special 'Juniper' line up at Washwood Heath on 14 April 2000 featuring Classes 334. 460 and 458. *Brian Morrison*

Back cover top: Bulleid's 'Leader' No 36001 pictured at Lewes on 31 August 1949. *C. C. B. Herbert*

Back cover centre: Class 313 No 313029 at King's Cross on 9 February 1978. *Dr L. A. Nixon*

Back cover bottom: The only surviving member of the 'Clayton' Bo-Bo locomotives, No 8568, awaiting preservation in July 1975, pictured near Hemel Hempstead. *R. G. Giddens*

The views expressed in this volume are those of the author and not necessarily those of the Publisher.

First published 2003

ISBN 0 7110 2836 2

Published by Ian Allan Publishing

an imprint of Ian Allan Publishing Ltd, Hersham, Surrey KT12 4RG.

Printed by Ian Allan Printing Ltd, Hersham, Surrey KT12 4RG.

Code: 0309/B3

Contents

Introduction

This is a history of railway blunders. After the initial, superb blunder of considering railways as a superior form of turnpike road and the national blunder of allowing the creation of too many railways too quickly and then letting them compete disastrously with one another, the Victorians lost their nerve and turned socialist, allowing the formation of monolithic, monopolistic railways, organised under central control from the top down. The 'monolithic' structure was developed by bitter experience as the most efficient way of operating a railway. On the run-up to Privatisation in 1992-4 the word 'monolithic' was used by the privatisers as a term of abuse, thus betraying their deep ignorance of railway matters. They allowed them their own track and trains and even permitted them to build their own locomotives, carriages and signalling if they so desired. Many companies did, and one thinks of such miserable examples of monolithic monopoly as the Great Western Railway and the London & North Western; the quality of blunders made by these shabby concerns has become a legend in our more efficient times. There were blunders, of course: in the 1830s and 1840s I. K. Brunel made some really brilliant blunders — right up to today's standards — but he was a genius. Other old blunders which have a modern flavour are those which arose from an urge to make money and neglect safety. Laws had to be passed to improve safety, and gradually the railways developed a notion known as 'public service'. This became the 'culture' of the railways in the latter part of the 19th century — the age of Victorian values. After nationalisation the Victorian ideal of solid engineering, service and safety was reinforced, and it was the abolition of this 'culture' which was specifically given as a reason for returning railways to their chaotic free-market roots.

The monolithic railway companies carried the traffic of two World Wars using a positively rigid 'vertically integrated' organisation to carry the heaviest traffic ever seen, using fewer engines and men than at any time in their history, and despite the best efforts of Nazi air power. After the second of these upheavals they were short of men, money and materials, but, because of the monolithic nature of their organisation, they were able to answer all calls on their resources with monotonous, taken-for-granted regularity: vast crowds on holiday, vast crowds daily to work, specials for football fanatics — all trains supplied on request. But railways were still making their own equipment, which prevented private enterprise making a profit out of public money, and, of course, everyone who travelled on a train was not buying petrol or using tyres or roads, so road-building contractors and tyre manufacturers were losing a potential goldmine. Bank loans on a gigantic scale were not being taken up. Petrol companies, tyre manufacturers, the road builders, banks — all monolithic, but in a nice way — had to be pampered.

In 1994 a kindly government returned railways to their roots. The boring old railway culture, developed on military lines, based on Victorian values of hoary experience, engineering expertise and public service, crowned by a total regard for safety, was replaced by 'privately owned' organisations. To be 'privately owned' would, we were promised in Parliament by the Government, convert the miserable British railway system into 'a world-class railway'. All that was needed to make trains run on time — cheaply — was seriously profit-minded men. No Marxist could have been more dogmatic. So the track signalling, stations and lots of development land was sold to an independent company at a bargain-basement price, and 25 Train Operating Companies (TOCs) were allowed to run the trains which they leased from one (or more) of three rolling-stock companies (RoSCos). Maintenance was contracted out and sub-contracted. The TOCs were in receipt of larger sums of public money than had been given to BR, but they promised that, at a later time in their franchise, they would pay the Government a rent for their franchise. So was brought into railway operating that vital ingredient which had been so sadly lacking during World War 2 — the 'discipline of the market place'. As is well known, market places are where calm engineers and self-effacing public servants carry out their duties and where profit is the last thing on anyone's mind. One could count on the fingers of one hand the TOCs which have paid anything back to the public purse as agreed, but one would require a computer to work out the hundreds of millions that have been paid to them to run the trains and to support their profit when the inefficiencies of the system caused them otherwise to make a loss. However, the new regime has raised the status of the rail traveller from passenger to customer, increased the production rate of blunders and raised their quality to previously unheard-of heights. They are now able to disrupt our railway system with a perfection previously achieved only by the Luftwaffe.

Lest I be accused of exaggerating, here are some recent examples of market-place efficiency in blundering. On 23 January 2003 Anglia Railways' 05.05 Norwich–Liverpool Street was hauled by one of the newer electric locomotives, No 90048. The train was scheduled to pass Shenfield, 20 miles from the London terminus, at 06.30.30. Shortly after this the pantograph dropped onto the roof of the engine because the valve retaining air pressure holding it up to the wire had failed. The train came to a stand on the up main at Gidea Park at about 06.40. It was vital to ascertain whether the failure had been caused by a defect in the overhead catenary or by a fault on the locomotive; if the former, Network Rail would be to blame and Anglia would not have to pay for the delays. Because of fragmentation, blame must be laid and fines paid to the injured parties. Time was thus lost whilst the wires were examined and the blame laid on the Class 90.

In the bad old days of monolithic ownership an engine would, without any doubt, have been found promptly from a nearby shunting yard or goods loop, and within minutes the train would have continued its journey. But in the more efficient and disciplined atmosphere of free-market railways, Anglia Railways does not own any engines except one designated rescue engine stationed at Colchester, 38 miles from Shenfield. This engine, No 47714, was despatched to the scene, taking its place in the queue of First Great Eastern commuter trains from Clacton, Colchester and Southend. All these were being funnelled along the up slow line, and delays to passengers and financial penalties to Anglia were mounting rapidly. Eventually No 47714 came alongside and passed the failure and was brought to a stand clear of the crossover from up slow to up main so that it could set back onto No 90048. But then the points could not be moved, because the track circuit recently occupied by the '47' had failed to clear and was showing 'occupied'. In the days of mechanical signalboxes overlooking the site, the signalman, having ascertained that the line was indeed quite clear, would have broken the seal on the 'Emergency release' plunger and operated it for that set of points. This apparently could not be done under the modern signalling system. The 'rescue engine' was sent on out of the way and now, and *only* now, the Controller was entitled to commandeer a locomotive to clear the line. At about 07.50 an EWS Class 37 left Temple Mills for Gidea Park. The track circuit was put right at 09.00, and customer trains could again begin to creep towards Liverpool Street along the up slow line. The Class 37 took the failed train forward at 9.20 and arrived in Liverpool Street at 9.45. Then No 47714 backed on and some time later set off with the empty stock for Norwich. The men were very keen to do a good job and hammered the '47' up Brentwood Bank, so much so that a fire broke out in the exhaust pipes and the train came to a stand at Shenfield. The power in the catenary was then turned off, once again bring all trains to a stand, and the fire brigade attended. A Class 66 was eventually commandeered and used to push the train into Ingatestone loop before leaving; commandeered engines may be used only to clear the line, not to take a train home — they must return to their rightful owner/lessee as quickly as possible. That evening the failed train was moved to Norwich, occasioning more delays to other trains.

The constant failure of equipment suggests overwork and a shortage of maintenance staff as costs are cut in the name of efficiency, while the long-winded, cumbersome operating procedures are the result of the weird 'free enterprise' thinking now applied to railway work. Today the quality of railway blunders is perhaps higher than at any time since Brunel formed his 'decided opinion' on the superiority of the atmospheric railway over steam haulage. This has been a great triumph of careful thought and decisive action, for which the instigators should receive their full and just reward.

Adrian Vaughan
Fakenham
July 2003

Acknowledgements

I gratefully acknowledge the generous assistance of many friends, more expert than I, who took time out from their own busy days to answer my questions, by post and e-mail — Lord Bill Bradshaw, Mike Christensen, David Collins, Brian Druce, Ted Gibbins, Richard Hardy, Brian Hart, Jim Low, Colin Marsden, Derek Milby, Kieth Montague, Brian Morrison, John Randell. I should also like to thank Robert Thomas and Keith Moore, librarians at the Institutions of Civil and Mechanical Engineers respectively, Mr Bryn Morgan of the House of Commons Information Office and, last but by no means least, Peter Waller at Ian Allan, for researching the photos and writing most of the captions, and Paul Cripps, for all his hard work in editing.

The Invention of the Locomotive

'George Stephenson invented the steam locomotive.' 'George Stephenson invented the *Rocket*.' These are popular beliefs — so popular that George's portrait is (or was) on the £5 note. But the popular belief is incorrect. In 1770 — 11 years before Stephenson's birth — Frenchman Nicholas Cugnot invented a twin-cylinder road-borne steam locomotive, used by the French army for hauling guns. The machine was difficult to steer, and this rectifiable defect caused the abandonment of a potentially world-beating invention.

The first rail-borne steam locomotive was designed in 1801 by Cornish mining engineer Richard Trevithick and built in 1803 at the Merthyr works of Samuel Homfray. At the time, George Stephenson was a fireman on a beam engine in Northumberland. In 1803 Trevithick's engine began hauling wagons on the steep, nine-mile-long Pen-y-darren Tramway from Homfray's works in Merthyr to the canal at Abercynon. Trevithick's locomotive was a curious mixture of innovation and blunder: it had only one cylinder — a retrograde step, in view of Cugnot's twin-cylinder machine of 30 years earlier — and Trevithick envisaged it not specifically as a locomotive engine but as a portable, to move itself to different sites of work, to drive pumps or other machinery, or to haul wagons; the locomotive moved by the adhesion between wheel and rail. Homfray made a great blunder in discarding the engine as a locomotive because it broke his tramplates. He put it out of sight, down a mine, where it worked for years as a pumping engine. Trevithick was disgusted at the lack of interest shown in his engine and made a great blunder by discarding the idea and going to South America. William James, a canal owner from Stratford-upon-Avon, saw the engine working on the tramway and realised its potential. He suggested a national trunk network of rail-roads for the steam locomotive. He was ignored.

In 1811 John Blenkinsop of Leeds and Timothy Hackworth of Wylam each built twin-cylinder machines. Each engine had one central flue through the boiler, with the fire grate in the flue. The Hackworth engine, for the Wylam colliery, was the first to mount its boiler and cylinders on a frame or chassis; its cylinders were placed vertically, resting on the frame, on each side of the boiler, at the footplate end. Blenkinsop's engine had no frames;

Stephenson did not invent the steam locomotive; static engines had existed for more than a century before he became involved in the railway industry, and, in France, Nicholas Cugnot had developed a steam-powered car, illustrated in model form in this photograph. *Ian Allan Library*

Richard Trevithick can perhaps be regarded as the true father of the steam locomotive, courtesy of his Pen-y-darren locomotive of 1804. Unfortunately, the original locomotive does not survive, but this replica, now on display at the Welsh Industrial and Maritime Museum, does. On 23 July 1981 it was displayed alongside a more modern version, when Class 56 No 56037 was officially named *Richard Trevithick*. *Richard Hooper*

its cylinders were mounted close together within the boiler, along its centre line. On each engine the cylinders drove a cogged shaft mounted centrally under the boiler; Hackworth's shaft drove the axles through a train of gears, while Blenkinsop's cogged wheel engaged in lugs cast into the side of the rails — it was a rack locomotive. In 1812, Hackworth built the world's first eight-wheeled locomotive, the boiler being carried on two four-wheeled bogies; drive was still by means of gears.

George Stephenson produced his first locomotive — the *Blucher* — in 1814, for the Killingworth colliery railway, but this was merely a combination of Blenkinsop's frameless boiler and cylinder layout and Hackworth's drive. In 1816 he brought out the *Locomotion*, another four-wheeled engine very similar to the *Blucher*, with a single-flue boiler. The cylinders were located within the boiler but were now in line with the wheels. A connecting road from each cylinder drove a pair of wheels. Each axle had a cog at its centre, under the boiler, and the two axles were then coupled by a chain around the sprockets. Stephenson had designed an excellent route for the Stockton & Darlington Railway, and for the opening of the new line in 1825 a new engine from Stephenson's works was to be used. The design was unchanged from 1816:

its wheels were cast-iron (without tyres) and were chain-coupled, it had no springs, and its boiler was so weak that it could not pull eight trucks; before it could be used on the opening train it had to have its boiler fitted with a U-shaped (or double) flue. It was also fitted with Hackworth's patent wheels, with cast-iron centres and shrunk-on wrought-iron tyres, and these were coupled with a connecting rod, such that, in drawings, the engine appears quite respectable, even though it was not.

In 1827 Stephenson's engines were still under threat from horse haulage, and, to provide the Stockton & Darlington with engines worthy of the name, Hackworth designed the *Royal George*. The world's first six-coupled engine, this had a double flue, to double the heating surface; it was also the first to run from new on cast-iron wheels with shrunk-on wrought-iron tyres and the first to

Arguably the most famous of all the early steam locomotives, Stephenson's *Rocket* won the Rainhill Trials on the Liverpool & Manchester Railway and established George and his son, Robert, as the foremost railway engineers of their time. In addition to acting as civil engineers on numerous projects, such as the London & Birmingham, they also established a locomotive-building company that bore their name for more than a century. *Ian Allan Library*

Locomotion — now preserved — of the Stockton & Darlington Railway was Stephenson's first conspicuous success as a locomotive engineer. Its origins as a design based upon existing practice in the coalfields of Tyneside is all too evident. The locomotive as preserved, however, is significantly different from that originally designed by Stephenson, having been modified by Timothy Hackworth with new boiler, wheels and coupling rods and by the replacement of Stephenson's chain-coupled wheels. Note also the two exhaust pipes in the chimney, exhausting steam up each inner wall rather than concentrically, which is ineffective in terms of drawing the fire. *Ian Allan Library*

have a feed-water heater, an axle-driven water-pump to put hot water into the boiler, a spring-loaded safety valve, self-lubricating bearings — and a cone (mounted concentrically at the base of the chimney) through which exhaust steam from the cylinders had to pass, forming a jet which rushed upwards, drawing air behind it to create a partial vacuum and so draw the fire fiercely and raise steam very quickly. This was the 'blast pipe' — one of the most important inventions in the history of locomotive design.

The *Rocket*

The *Rocket* of 1829 was a radical redevelopment of *Lancashire Witch*, built at the locomotive works of Robert Stephenson & Co in Newcastle in 1828. This four-wheeled engine had inclined outside cylinders at the rear, driving the front wheels. George Stephenson's useless, bellows-driven air blast to the fire was discarded in favour of Hackworth's blastpipe. Externally, the boiler and cylinders of the *Rocket* looked like those of the *Lancashire Witch*, but the engine was profoundly different where it mattered — on the inside. George Stephenson's heating flue was removed in favour of a multi-tubular heating element leading from a water-jacketed firebox attached to one end of the boiler. The blastpipe and the multi-tubular firebox together were the key to vastly increased steam production and thus to greater power. Neither was the invention of the Stephensons: the blastpipe was Hackworth's, and the multi-tubular firebox was the suggestion of Henry Booth, Secretary of the Liverpool & Manchester Railway Co. The myth of George Stephenson as the inventor of the *Rocket* — or even as an innovator in locomotive design — is a blunder of history long overdue for rectification.

Parliamentary Blunders

Public railways began with the Surrey Iron Railway, opened in 1805. This was an extension of the turnpike road principle: the company was authorised by Act of Parliament to build the road, but by the same law anyone could make use of it — it was a public rail-road. It was a double track, on which haulage was entirely by animals.

The next public railway to open in Britain was the Stockton & Darlington Railway (SDR), opened in 1825. The S&DR Directors blundered in constructing their railway as a single track with passing loops. Collieries and passenger coach operators ran their trains along the line as they pleased, without a timetable, without signalling. Broken-down steam engines blocked the single track for horse-drawn trains, while horse-drawn trains got in the way of locomotives with a good head of steam. There were daily confrontations between opposing trains as to who should reverse to the passing loop. S&DR management realised that, for the sake of efficiency, the good old British free-for-all could not be allowed on railways. Consequently it got rid of private operators, took over traffic operating and developed a rule book. Once this had been done, from September 1833, the company was able to operate its first timetabled, locomotive-hauled, passenger service. From the outset, railway operation demanded centralised management, control, rules and regulations such as had previously existed only in the armed services.

The Liverpool & Manchester Railway (LMR) opened on 15 September 1830. It was a double-track line and was a public, toll, rail-road. A few colliery owners tried to run their own trains, but it soon became obvious that efficient railway working required central control of operations, and private operation died out. Those who were not railway managers (then as now) condemned this as a *monopoly* — which was a *very bad word*. The true nature of railways developed as they grew bigger and older, just as if a railway were a person.

Railways are socialistic entities — even though they were formed with private money. A truly national railway service had to be under one management — or the fewest number of managements — so as to render the service continuous from end to end. The rules governing operations and equipment had to be the same (or at least compatible) throughout all management's jurisdictions. These truths came as a surprise even to the parents of the railways — the investors. A Parliamentary Select Committee investigated the new phenomenon of railways in 1839 and concluded that 'railways are a natural monopoly'. This was an alarming discovery, and there were two schools of thought on how to deal with it. A minority, including Robert Stephenson, Thomas Brassey, William Gladstone and the Earl of Dalhousie, believed it would be best if individual investors provided the capital to construct such railways as the *Government* considered essential. This would avoid unnecessary expenses in competition between companies — inside and outside Parliament. Gladstone believed that railways ought to be a state enterprise, like the newly formed Post Office. Robert Stephenson and Thomas Brassey wanted state control and private money so that investors would always obtain their 10% and the nation a rational network of railways. This was opposed by the majority, because a calm market in railway promotions, shares and loans is not a hugely profitable one for lawyers, bankers, and share speculators. Thousands of miles of unprofitable railway were built — many competing, many just to trade in their shares — under the pretext that many railways made for virtuous competition as the antidote to railway 'monopoly'. Having allowed all these railways to be built, Parliament then forced them all, by law, to act as public services. Railway profits and share values were gradually crippled by financial pressure simultaneously from the free-market and public-service directions. This inability of Parliament to decide whether railways were private-enterprise or public-service bodies was an enormous blunder which undermined and weakened the companies.

Railway companies could only be formed by Act of Parliament because of their need for 'joint stock company' status and for the compulsory purchase of land. To gain its Act of Incorporation a private railway company had to prove that building a railway would be in the public interest. The promoters of the GWR in 1835 spent

Initially the promoters of railways believed that, once their line had been constructed, open access would allow anyone to use the line in a similar way to the use of contemporary roads. This spirit was reflected in the early Acts, but the practicalities of open access in an unregulated environment soon proved that such a policy was impossible, and the early railways, such as the Stockton & Darlington and the Liverpool & Manchester, soon abandoned the idea in favour of running their own services. This early print records the Liverpool & Manchester shortly after its opening and shows well the rudimentary facilities provided at early stations. The location, Parkside, was where William Huskisson was killed — the first railway fatality. *Ian Allan Library*

approximately £75,000 — in gold sovereigns — to prove this obvious fact: that it would be a public benefit to build a railway from Bristol to London. Then they had to compensate the owners of land for being forced by law to sell to the company, which thus paid twice for the land.

The Eastern Counties Railway proved that it was in the public interest to build it and was obliged by law to spend £370,000, on legal fees to get its Act, land purchase and compensation to owners. It had a capital of £1.6 million yet was financially crippled even before it started construction. Samuel Smiles, Company Secretary of the South Eastern Railway, lambasted Parliament for allowing the railway speculators to create the 1844-7 'mania', referring to 'a tissue of legislative bungling involving enormous loss to the Nation. The want of foresight displayed by both Houses in obstructing the railway system so long as it was based on sound commercial principles was equalled only by the fatal facility with which they granted projects based on the wildest speculation.'

Robert Stephenson, in his Inaugural Presidential speech to the Institution of Civil Engineers, in January 1856, hurled scorn on the huge cost of Parliamentary procedures necessary to obtain an Act of Incorporation even when the railway concerned was of the most vital national importance. He cited the example of the Trent Valley Railway, from Rugby to Stafford, intended to shorten the route from London to Holyhead by removing the detour westwards through Birmingham and Wolverhampton. 'The policy of Parliament would seem to be to put the public to expense to make costs for lawyers and fees for Officers of Parliament. Is it possible to conceive anything more monstrous than to condemn 19 parties to the same contentious litigation? They each and all had to bear the costs of opposing all other Bills for the same railway. The ingenuity of man could scarcely devise a more costly system of obtaining a railway Act.' The cost of obtaining the Trent Valley Act was £630,000 — more than it would cost to build the line.

Railways have always been interfered with by politicians, like no other industry. All railway companies had their maximum charges set by Parliament, and the Liverpool & Manchester Railway was obliged by its Act to reduce its charges if it made more than 10% profit. In 1842 Parliament laid a tax on all railway tickets but not on road fares. The 1842 Act also obliged the railways to carry military troop traffic, at a price agreeable to the Government. But the Government invested no money in railways, instead taking actions that inflated their costs. The Regulation of Railways Act 1844 laid some severe social responsibilities on the companies: any railway paying 10% dividend had to reduce its fares and freight charges; they were also forced to carry the military and the poor at one penny a mile — laudable but not profitable. In 1854 an Act was passed forcing railway companies to build whatever facilities a trader might request, while in 1873 a further Act obliged the railways to make available for public inspection, at every station, all their charges. No other trader ever had to do that, and, of course, rival hauliers made good use of the Act, especially with the arrival of the motor bus and lorry. In 1896 yet another Act made it very difficult for railway companies to raise — or lower — their fares.

At the outbreak of World War 1 the Government took control of railways and guaranteed the companies their 1913 income. For that money they carried all traffic, civilian and military — a weight of traffic far greater than that of 1913. All the railway companies suffered from lack of maintenance, and the North British Railway and the South Eastern & Chatham were brought to the verge of bankruptcy by the cost of hauling so much traffic for no financial reward.

In the Railway Act 1921 the Government handed back the railways — worn out, lacking in funds to repair themselves — to their owners, without adequate compensation for all the damage caused. The level at which the Act fixed railway charges was supposed to be enough to guarantee a certain net income, but no other trader was restricted by law in to what he could earn. The Act perpetuated the obligation on railways to make available to the public their freight-carrying charges, but no other haulier was made to disclose his rates. Thus a road haulier could go to a factory and immediately quote a rate below the railway rate; if the railway tried to do the same, it had to spend weeks waiting for a tribunal to discuss the matter. Road hauliers were under no trading restrictions, nor (unlike railways) did they have to provide humane working conditions for their drivers — and all the while the railway companies, through the rating system, were contributing towards the cost of maintaining the roads.

Governments insisted that the railways were private enterprise but used them to further Government economic policy. Road haulage was never subjected to that abuse. The tax on railway tickets was finally abolished in 1929, when the Government diverted this tax income to the relief of unemployment: what the railways saved in ticket tax remission they had to spend on labour-intensive improvements to their infrastructure.

Railways were always the largest ratepayer in every parish they passed through. The method of calculating the contribution due from each occupier in a parish had been laid down in Elizabethan times. The railway companies brought the matter before Parliament on several occasions, but Honourable Members saw no reason to untether an extremely valuable milch-cow. Typically, 80% of the money raised by parish rates came from the railway that passed through its fields. The GWR's annual contribution to parish rates in 1928 was £1.5 million, equivalent to $1\frac{1}{2}$% extra on the dividend — which increase would have made people more interested in buying railway shares. Much of the parish rates was spent on the maintenance of roads. The rating burden of a road freight or passenger-haulage company was minuscule by comparison, limited only to the garage in which the vehicle(s) was housed. In 1929 Parliament allowed the railways parish-rate relief — but only on condition that the money be spent on projects to relieve unemployment.

Throughout the 1930s even Chambers of Trade, on behalf of the railway companies, lobbied Parliament (unsuccessfully) to remove outdated laws based on fear of a railway monopoly that no longer existed and to allow railway companies the same commercial freedoms as road hauliers.

Competition

Perhaps the greatest blunder in the early history of Britain's railways occurred when Parliament allowed the establishment of a national transport system to become a free-for-all, a share-dealing racket, a happy hunting-ground for lawyers and engineers looking for a job, rather than a planned series of trunk railways. If the French government had pioneered railways they would have been built with private and state capital and under state direction, and competition would have been avoided — because that is how things were done in France. It was orderly and slow but made a safe investment and produced a slow and orderly railway. With that orderliness for our guide, we in Britain would still have built ours through individual enterprise, because that is how we do things here. In 1824 Charles Sylvester wrote a report for the Liverpool & Manchester Railway Committee in which he concluded that such a route could not fail to be extremely profitable but warned against 'a rage for proposing new rail-roads and the delusion that because rail-roads are better than canal or high roads, they will answer everywhere. The pretensions held out by some projectors do not appear to be warranted. This new application of locomotive power is of infinite importance to the country and I should regret to see it abused.'

Of course, it was abused. In a private-enterprise situation, how could it be otherwise? Private enterprise created railways, and competition between the various entrepreneurs reduced their railways' profitability. Only those who do not understand railway history can have an enthusiasm for competition on railways. A very large part of the British railway system was built as result of strategic manœuvring rather than for sound commercial reasons. The GWR, for instance, involved itself in building railways to Weymouth in an attempt to prevent the LSWR entering the West Country. Such competition between companies for routes resulted in railways' being built before the companies could really afford them. This additional capital was either unproductive or not very productive but had to be spent to 'defend' territory. Large sums of

'A Locomotive Kidnapped' was the dramatic heading given in Grinling's classic history of the Great Northern Railway to the story of the relationship between the GNR and MR at Nottingham. Whilst the identity of the locomotive 'kidnapped' is unknown, it is possible that it was one of a batch of 50 2-2-2s (Nos 1-50) constructed by Sharp between 1847 and 1850 for passenger work. The majority of the class, as evinced by No 45 shown here, were subsequently converted into tank locomotives. Despite the modifications, the photograph gives a good impression of how an early GNR locomotive would have appeared. *Ian Allan Library*

money were lost from the profits of the railway companies through 'wars' among themselves conducted through the courts, in Parliament and by the construction of 'aggressive' or 'defensive' railways. This is the free-enterprise way. It is very wasteful.

In February 1848, George Carr Glynn — no minor figure in the Victorian world, being Chairman of Glynn's Bank and of the LNWR — told his shareholders: 'If, instead of trying to destroy each other's interests, railway companies can be brought to unite — if, instead of encouraging competition, Parliament will impose on all railway companies a proper system of fares and charges so as to secure the public welfare — then railways will continue to be a safe investment.' But business is business, and the railway companies had no choice — in a free market — but to try and destroy each other. Lack of profit prevented the installation of proper brakes and signalling.

July 1850 saw the opening of the Ambergate, Nottingham, Boston & Eastern Junction Railway — a line from a junction with the GNR near Grantham to a junction with the Midland Railway at Colwick, four miles east of the MR's Nottingham station. The MR agreed to Ambergate trains' using its Nottingham station. By 1852 the Midland had, for a couple of years, been talking with the LNWR for amalgamation into a single company. These two companies owned the only route from London to Leeds and York, via Euston, Rugby and Derby, and between them controlled most of the railways of England north of the Thames. In 1852 the imminent completion of the GNR's 'Towns Line' — creating the East Coast main line and cutting an hour off the route from London to York and Scotland — prompted the MR to open talks with the Ambergate with a view to taking over the latter. A large shareholder in the GNR, Mr Graham Hutchinson, gained

a controlling interest in the Ambergate and used this to prevent its sale to the Midland; he then established a working agreement between the Ambergate and the GNR, whereupon the Midland paid a dissatisfied GNR shareholder to go to court to object! This objection was duly upheld, and the GN-Ambergate working agreement could not be implemented.

The GNR's main line from Peterborough to Doncaster opened on 1 August 1852. A connecting service from Grantham to Nottingham was operated by the Ambergate company — but drawn by a GNR engine on hire to the Ambergate. When the bright-green GNR engine rolled into Midland-maroon Nottingham, local officials were outraged, believing a GN train had illegally entered their station under cover of the MR-Ambergate agreement: at once they backed a Midland engine onto the GN engine, the carriages were drawn off the rear and another engine was put in behind. Despite all protestations from the Ambergate driver that his company was only borrowing the GN engine (and in spite of his best endeavours with the regulator!) he and his steed were dragged away to a shed and thrust rudely inside. The doors were locked and the rails leading to the shed taken away, and it would take seven months for the GNR to get its engine back.

At this stage Edmond Denison, Chairman of the GNR, suggested to Edward Ellis, Chairman of the Midland Railway, that the two companies should amalgamate. The content of Denison's letter can be inferred from Ellis's reply:

'Our Board is equally alive to the evils of competition; the needless expenditure in running duplicate trains and the still more imperative necessity of preventing a reckless outlay on the construction of new lines [but] you are aware that for many years the L&NWR and MR

The Portsmouth Direct line from Guildford to Havant was constructed by contractor Thomas Brassey as a speculative venture threatening both the LSWR and LBSCR. In the event, it was the LSWR that ended up in control of the route. On 28 October 1962 the down 'Bournemouth Belle', diverted via the route as a result of engineering work at Wootton and Swaythling, passes Petersfield behind 'West Country' Pacific No 34010 *Sidmouth*. Unusually for a speculatively constructed line, the Portsmouth Direct route has become the primary line between London, and this part of Hampshire and has been electrified for almost 70 years. Perhaps Brassey was right after all? *J. C. Haydon*

companies have cultivated an intimate alliance and negotiations are on foot for a closer union.'

Clearly, as early as 1852 the GNR, MR and LNWR realised the need for a truly national railway without expensive competition. In that year the MR and LNWR placed a Bill before Parliament for amalgamation and — jointly — for a railway from Leicester to the GNR at Hitchin. Parliament refused the LNWR/MR Amalgamation Bill on the grounds of maintaining competition, whereupon the GNR put in a Bill to build a railway from the main line at Sandy to Bedford, across the path of the Leicester–Hitchin, which Bill was passed.

The other great problem for the 19th-century railway

Competing routes could often arise as a result of a dispute between existing companies. For the Midland Railway access to Carlisle was only possible via the line north from Clapham via Sedbergh to Low Gill. South of Ingleton, the MR controlled the line, but north of Ingleton, through to Carlisle, the route belonged to its fierce rival, the LNWR, and the MR was forced to rely upon running powers over the route. By the 1860s, competition between these Leviathans was such that the LNWR caused serious problems for MR services trying to reach Carlisle. As a result, the MR promoted its own competing route to Carlisle — the Settle & Carlisle — which opened in 1876. Thus there were now three routes running through the sparsely populated area of Cumbria, and inevitably there was to be a casualty, with the route via Sedbergh succumbing; indeed, for much of the last quarter of the 20th century, even the S&C was not secure. Typifying the scenery through which much of both the Sedbergh and S&C routes passed is this portrait of '9F' No 92012 with a limestone train at Ais Gill Summit on 25 August 1966. No doubt there are plenty of sheep, but fare-paying passengers are thin on the ground in this moorland environment. *Alan L. Bailey*

companies was railways built by contractors. Sir Richard Moon, the granite-faced Chairman of the LNWR, said of them in 1862: 'No proprietors [*ie* ordinary shareholders] are willing to come forward to make a railway. Not one of the great companies can raise sixpence except by guaranteed interest. Railways are made by contractors, engineers and speculators who live on the fears of the companies.' A contractor would build a railway linking two opposing companies; neither company would want the line, but one of them had to buy it to keep the other out. These railways, known as 'contractor's lines', amounted to nothing less than blackmail — except that their construction was entirely legal. The effect on the established companies of being obliged to spend huge sums of money to buy railways they did not actually want is obvious.

Between London and Portsmouth two railway companies — the LSWR and the LBSCR — provided transport; the distance by either route was 95 miles, and the companies had, after bitter experience of fare-cutting competition, reached an agreement to pool and share the combined revenues of the Portsmouth traffic. However, between 1853 and 1857 the contractor Thomas Brassey built the Portsmouth Direct Railway (PDR) from the LSWR at Godalming, through Rowlands Castle, to the LBSCR at Havant, on the Brighton–Portsmouth route. The PDR took 20 miles off the existing rail routes and was purely speculative — forcing either the LBSCR or LSWR to buy it. It had to be built cheaply so as to be sold for the greatest profit. The country through which it ran was exceedingly hilly, so the 'undulating' principle of construction, with very steep gradients, was adopted to

Another corridor to witness three competing lines was the Leen Valley, to the north of Nottingham. Here the Midland, Great Northern and, from 1892, Great Central railways operated lines that were but a stone's throw apart. Whilst the competition engendered was undoubtedly useful for the people who lived in the district and for the owners of the local coal mines, come Grouping and Nationalisation, this unnecessary duplication was wasteful and it was inevitable that lines would go. In the event, the Leen Valley suffered worse than Cumbria, in as much as all three routes lost their passenger services and only one route remained to serve the surviving coal mines. However, a through service — known as the Robin Hood line — has now been reinstated. Here, on 12 June 1957, the down 'South Yorkshireman' crosses Bulwell Viaduct, to the north of Nottingham, on the erstwhile GCR, behind 'Black Five' No 45219. *P. J. Lynch*

avoid expensive earthworks, the excuse being that speed downhill would compensate for crawling uphill. In spite of this there were still some very heavy earthworks, including the half-mile-long Buriton Tunnel. Speed (in steam days, at any rate) could never be high downhill because of the sharp curves snaking through the hills and in particular around Rowlands Castle. Miles of it were at 1 in 100, and there was a three-mile incline at 1 in 80. Whoever bought it would be lumbered with a thunderingly expensive line to operate. Furthermore, after the expense of buying it, the new owner would possess a line which competed with its existing route, because the PDR was shorter than either existing line. Therefore, purely because of mileage, fares between London and Portsmouth would have to be reduced — not a pretty prospect for shareholders after they have just paid a lot of money for an expensive-to-work railway line.

Neither the LSWR nor the LBSCR wanted to play, so Brassey went to Parliament and in 1854 obtained an Act which was the equivalent of a highwayman's pistol held at the head of a coach driver. He obtained authority for a railway from the PDR at Godalming to the South Eastern Railway at Shalford, between Guildford and Redhill. Brassey then found the SER in a pious mood, refusing to break its non-agression pact with the LSWR. However, the LSWR did not trust the SER to remain pious forever and, breaking its promise to the SER, entered into an agreement with Brassey. In 1858, with the PDR ready for use — and the extension complete except for track — the LSWR agreed to pay Brassey £18,000 a year for the privilege of running trains over this hilly, corkscrew railway. That is as near highway robbery as one can get without actually committing the crime. At this juncture, Brassey threw away his 'pistol', obtaining an Act allowing him to abandon the Godalming–Shalford line.

The LBSCR decided that the new route was a threat to its traffic to Portsmouth and decided to oppose the LSWR in that area. The LSWR was aware of the hostility and, intending to begin its service on 1 January 1859, ran a trial train loaded with fighting navvies to test the reaction of their erstwhile friends. At Havant the train was confronted with an engine chained to the track and a platoon of LBSCR men surrounding it. The LSWR 'bruisers' descended from their train, and, while some attacked the garrison, others unchained the engine and called their engine across the junction. A battle between engines took place, forcing the LBSCR engine backwards. To the rear of the fight, rails were removed. Fighting ensued to replace the rails. The Battle of Havant ended in the withdrawal of the LSWR, and both companies continued their quarrel in court. The learned judge found in favour of the LSWR, whereupon the LBSCR began a price war on the Portsmouth run and even reduced its fares on the Southampton–Ryde ferry service. All this achieved was a heavy financial loss on the operations, and when each side had had enough punishment they reverted to their previous non-agression pact, pooling and sharing their Portsmouth revenues.

Thomas Brassey's friend and business partner, Samuel Morton Peto, built the East Suffolk Railway for sale to the Eastern Counties (ECR). The East Suffolk ran from Great

The story of the Oxford, Worcester & Wolverhampton Railway is a classic in the annals of the machinations between the early railway companies. However, despite being constructed to standard (4ft 8½in) gauge, it fell within the orbit of the GWR, resulting in the construction of stations to the design of Brunel. Much of this heritage is now lost, but the station building at Charlbury — pictured on 19 June 1980 — remains.
G. Wright

Yarmouth and Lowestoft to an end-on junction with the Eastern Union Railway (EUR) at Woodbridge, a few miles from Ipswich. But, while the new line was under construction, Peto's collaborator on the ECR, the Chairman, David Waddington, had been forced by his shareholders to resign owing to his mismanagement and corruption in office, a great deal of which had been for the benefit of Peto. The latter then proposed a railway — which he called the Colchester, Maldon & Pitsea (CM&P) — with which he could blackmail the ECR: such a railway, from the ECR at Colchester to Peto's and Brassey's London–Tilbury line at Pitsea, could be used by the EUR to divert all Norwich, Great Yarmouth and Lowestoft traffic to London independently of the ECR line from Colchester. Peto engaged an engineer, plans were made, a prospectus of dubious factuality was published and investors wooed. A Bill was put before Parliament. At a public meeting in Colchester Town Hall on 24 December 1856 Peto was asked if he would proceed with the CM&P if the ECR agreed to purchase the East Suffolk. Peto promised that it would be built, and on the basis of that promise — and of Peto's great reputation as a Christian philanthropist — people paid cash for shares. The Pitsea Bill duly passed its second reading, the ECR directors gave in to Peto's blackmail and agreed to purchase the East Suffolk. Peto immediately withdrew the Pitsea Railway Bill from Parliament, and all shares in that company were worthless.

The Oxford, Worcester & Wolverhampton Railway (OWWR) was by law betrothed to the GWR; the latter company had contributed a great deal of the OWWR's capital and was the only railway empowered under the OWWR's Act to lease or purchase it. The Act also obliged the OWWR to lay the broad gauge. That was the law, but the OWWR was thrusting into the heart of the territories of the Midland Railway and LNWR. Under construction at the same time were the independent Shrewsbury & Chester and Shrewsbury & Birmingham railways, which were being subjected by severe bullying by the LNWR to force their respective shareholders to sell out; meanwhile the GWR was constructing its own main line from Oxford to Banbury, Leamington and Birmingham, which also threatened the LNWR and the MR. In due course the OWWR got into financial difficulties, and, with the GWR unable to supply more money, the LNWR, seizing its chance, sent Peto to offer the necessary cash in return for the contract to complete the railway and a seat for his creature — a perjured lawyer by the name of Parson — on the OWWR Board. Once installed, and with Peto behind him pulling the strings, Parson wasted huge amounts of OWWR shareholders' money in trying to sell the railway to the LNWR, in spite of the fact that this was entirely illegal and could never succeed. Peto also promoted more than one railway from Oxford to London as a threat to the GWR. For both companies the cost of these Parliamentary and legal battles was huge, and there was an additional cost — Peto's and Parson's ruthless drive to get their way cost the GWR Chairman, Charles Russell, his life. He suffered a nervous breakdown and retired from the company in August 1855 and shot himself the following May.

In 1858 the OWWR made peace with the GWR — which had been its friend all the time. Meanwhile the Shrewsbury & Chester and Shrewsbury & Birmingham, indignant at their treatment by the LNWR, sold out to the Great Western!

Blunders of Snobbery

Britain's first main-line trunk railway was the Liverpool & Manchester (L&MR), the idea for which came from William James, business partner of George Stephenson. James went bankrupt in 1824 as a result of his efforts to promote and survey the line, and George Stephenson took over as Engineer. Stephenson, uneducated and self-taught, had the granite determination necessary to build the line but lacked the mathematics and surveying skills to make the plans; he also spoke with a wild Northumbrian accent, and Parliamentary committees were just too snobbish to take him seriously, even when he spoke good sense. Asked what cows would do if they saw an engine with a red-hot chimney, he replied that they would think it was painted red. When he told the L&MR Committee that he would 'float' an embankment across Chat Moss on mats of brushwood, they were derisive, being ignorant of the fact that this was standard practice on swampy ground and that much of London's Thameside was founded on brushwood rafts. His plans were found to be flawed mathematically — although the idea was a good one — and the Bill for the L&MR was thrown out.

The L&MR Committee realised that, whatever George Stephenson's practical abilities, they would never get the Bill through Parliament with him as their spokesman. So they sacked George in favour of the famous (and educated) Rennie brothers — John and George. They agreed only to be the Consulting Engineers of the line — so as to qualify for their fee — but handed over the actual survey work to the elegant, educated and highly sophisticated Charles Vignoles. Vignoles felt obliged to alter Stephenson's route and took a more southerly line, bringing it more directly to Edgehill and thus making necessary what Stephenson would have avoided — the huge and expensive cutting at Olive Mount, two miles long and at one point 100ft deep. Vignoles — very much at ease with the gentlemen Members of Parliament, and in spite of the new-found expense of his line — easily talked the case for the railway through Parliament. The Liverpool & Manchester Railway Co was incorporated by Act of Parliament on 5 May 1826.

Manchester, second only to London as Britain's greatest centre of manufacture, had been successful in preventing the railway from entering the town, the terminus being out at Salford — a ridiculous blunder on the part of the Mancunians and one which was corrected at the extreme expense of the L&MR company, which had to fight Manchester in Parliament on three occasions before it got an Act permitting it to enter and serve the traders of that great town. These were the very people who would later complain of the great expense of railway charges!

Once the Act of 1826 had been obtained, the L&MR directors re-employed George Stephenson as Resident Engineer, whereupon the Rennies resigned in disgust. This was a great shame in one respect, at least: the Rennie brothers had recommended that the new railway be built with a gauge of 5ft 6in, but this progressive idea was ignored. George Stephenson hated the Rennies, who despised him, and, in spite of the merit in a wider gauge, George was not going to adopt their ideas, and the perfect gauge was lost. Only 375 miles of railway then existed, all of it parochial. If the L&MR — the first trunk main line — had been built to the 5ft 6in gauge, it seems likely that all other railways leading out of it to the north, south and east would also have been 5ft 6in, creating a usefully wide 'standard gauge'.

Work commenced on the Edgehill–Wapping Docks tunnel in October 1826. George Stephenson, with Vignoles as his assistant, had laid out the tunnel centre line on the ground above, and on that line shafts were dug down to the tunnel level. But sighting a centre line perfectly straight over the whole 1¼ miles was made very difficult because it passed through a built-up area with houses blocking the line of sight. Stephenson did not have sufficient skill under these circumstances and relied on Vignoles to put him right. Vignoles supervised the tunnelling and realised without alarm that the various headings were not perfectly in line; indeed, some were as much as 20ft off-centre and, had they continued unchecked, would have missed each other underground. There was a row between Vignoles, Stephenson and the Directors of the railway, following which Vignoles, who, as an ex-Army officer-engineer, resented playing second fiddle to the rough Northumbrian, left. Stephenson's chosen assistant, Joseph Locke, was brought in to correct the line of separate headings — some task — but Locke completed it successfully and gained great credit with the L&MR directors. This, of course, did not please Stephenson, who remained insecure at his own lack of education.

The Liverpool & Manchester was opened, inconveniently far away from the centre of Liverpool, on 15 September 1830. Soon a new 1¼-mile stretch of line down into the town (at Lime Street), requiring a tunnel, was proposed. Stephenson's inaccuracies of measurement soon became apparent, the sections of tunnel being well out of line. Joseph Locke was again called in, and again he sorted out the mess to perfect accuracy. By now George Stephenson was getting very sick of Locke. His resentment of his pupil developed as the mileage they were building increased until it was outright hatred on Stephenson's part and real sorrow at the dissolution of their friendship on Locke's part. Stephenson's jealousy lost him the services of one of the three great engineers of the classical period of British railway construction.

George Stephenson grew up with horse-drawn colliery tramways where the rails were nailed to stone blocks. Stone blocks were notorious for sinking into the ground under the weight of a loaded wagon and would sink even more under the weight of a steam locomotive. Neither George nor his son Robert saw any reason to alter this method when building trunk railways for locomotive-hauled trains. Nor did most of the engineering establishment of the day, but two men — I. K. Brunel and Joseph Locke — realised the need for improvement, and it was Locke who developed the track which remained the standard on Britain's railways for 100 years. George Stephenson had used timber cross-sleepers to carry the rails over Chat Moss where the ground was too soft for stone blocks and the long sleepers spread the weight of the train over a wide area. The sleepers could easily be under-packed to maintain their level. Travelling by train over Chat Moss, everyone noticed that the ride was softer and smoother than over the stone blocks of the rest of the line. But only Joseph Locke saw that cross-sleepered track was the practical way forward for all railways. All other experts dismissed that method of construction as fit only for temporary ways and for unique situations such as Chat Moss. The generally held belief, shared by Brunel, was that the hardest possible road bed should be provided for trains.

On the 82-mile-long Grand Junction Railway, from Birmingham to Warrington, Locke laid very heavy (84lb/yd) 'bullhead' rails resting in cast-iron 'chairs' spiked to timber cross-sleepers. Yet such was the prestige of George Stephenson — and the unquestioning acceptance of all that he gave forth — that Locke was obliged to replace 60 miles of cross-sleepers with stone blocks. Brunel saw the shortcomings of the Stephenson system but did not see the simplicity and ease of maintenance of Locke's cross-sleepers. Brunel designed a very expensive track, resting the rails on longitudinal sleepers pinned to the ground every 15ft by a massive post driven deep into the ground. The result was a roller-coaster ride and a lot of derailments. The GWR directors consoled themselves that Stephenson's stone-block track was even worse — but no-one wanted to notice the smooth running of Locke's 20 miles of cross-sleepers, and Nicholas Wood, the most celebrated 'outside' expert on railways of the time, stated that 'cross sleepers cannot be considered as a permanent description of road'.

Brunellian Blunders

There can be no doubt about the genius of Isambard Kingdom Brunel as a civil engineer, but equally there can be no doubt that he was a serial blunderer whose odd ideas wasted huge amounts of his shareholders' money.

Brunel decided to place his rails 7ft 0¼in apart, stating two objectives:

1. to reduce friction in axle bearings;
2. to lower the centre of gravity of the vehicle.

His thoughts were influenced by stage coaches, which had large wheels and the body within them for increased stability. In his carefully considered written argument to his directors, pleading the case for this 'broad gauge', Brunel noted that standard-gauge carriages had their bodies above the wheels, 'which raises the body unnecessarily high whilst restricting the size of the wheels'. He wrote that if a railway-carriage body lay between the wheels he could achieve a low centre of gravity, while wheels of 'unlimited size' would reduce friction between the axle and its bearing, because, he said, friction was reduced in proportion to the increase in diameter of the wheel. The standard wheel diameter was about 3ft 6in, while Brunel's carriage wheels were 4ft. The reduction in friction in the bearing was negligible. Not much gain to offset the disadvantage of a nationally incompatible gauge.

Only four passenger carriages were built with the wheels outside the body, because this was severely impractical — the width of the carriage floor could never be more than 6ft 9in. Brunel then put his carriage bodies over the wheels, in the manner of the despised standard gauge and thus losing the low centre of gravity he had said was so important. He retained his relatively large-diameter wheels quite simply, by bringing them through the carriage floor under metal cowlings beneath the seats, but he had abandoned his allegedly carefully thought-out reasons for installing the broad gauge; he could have had his relatively large-diameter wheels on standard-gauge track (which itself could have been 5ft 6in if the Rennies' inspired suggestion had been acted upon in 1828).

The broad-gauge track was retained and extended for

Now portrayed as one of the greatest Britons ever, courtesy of the BBC (and strong support from a certain university), Isambard Kingdom Brunel was an undoubted genius but one whose successes — the *Great Britain*, the Clifton suspension bridge, much of the Great Western Railway's early development, etc — can be overshadowed by his failures. This statue, located at Paddington, sees the great man looking out over one of his greatest creations but one which has thrived largely because it ultimately rejected the impractical foundations upon which it was constructed. *John Glover*

Right: One factor in Brunel's choice of the broad gauge was that it allowed for the carriage wheels to be outside the body of the coach — as with contemporary stage-coach practice — which would, therefore, reduce the centre of gravity and improve the quality of the ride. This view of an 1838 broad-gauge posting saloon illustrates one fundamental flaw in the design — the difficulty of access. *Ian Allan Library*

Below: Following the use of the outside wheels, Brunel's later coaching stock reverted to the more common practice of locating the wheels under the frames. This had the advantage of improving access but also countered the benefit that broad gauge had offered in terms of a lower centre of gravity. *Ian Allan Library*

20 years, even though the reasoning behind it had been shown to be flawed within a year of opening the first 20 miles of the line. Brunel increased the GWR's construction costs, created the inconvenience of transferring freight from gauge to gauge and ultimately was responsible for the considerable expense of laying a third rail and then of converting the broad gauge to that of the rest of the country.

Brunel's locomotives

Brunel did not design the original engines for the GWR — that was done by the manufacturers — but their designs had to conform to strict instructions laid down by him. These were that the engines should:

1. not weigh more than $10\frac{1}{2}$ tons;
2. be capable of a velocity of 30mph;
3. have a maximum piston speed of 280ft per minute;
4. have a maximum boiler pressure of 50psi.

Standard-gauge locomotives then weighed 14 tons and happily ran at 30mph with a piston speed double that required by Brunel. His specified boiler pressure was timidly low compared with the 90psi used by the very progressive locomotive engineer John Gray (who, incidentally, showed that Brunel's concern for a low centre of gravity was greatly exaggerated). Brunel's entirely unnecessary restriction on the speed of the piston forced the manufacturers to use a relatively short-stroke piston coupled to an over-large driving wheel. These big wheels increased the weight of the engine, and the boilers and cylinders had to be reduced in size to compensate. Even so, the resulting engines weighed between $14\frac{1}{2}$ and $18\frac{3}{4}$ tons. The fact that these freak locomotives had axles 7ft long could not compensate for the fact that they would always be short of steam and would have difficulty hauling even the lightest trains: *Vulcan* could manage only 21mph with 18 tons, and *Aeolus*, loaded with 43 tons, managed just 6.5mph for $2\frac{1}{2}$ miles before running out of steam. Brunel denied any responsibility

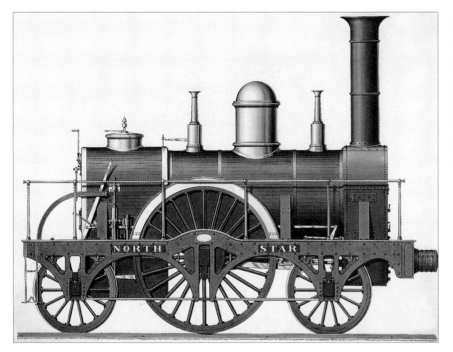

Although Brunel did not, himself, design locomotives, he did lay down stringent restrictions to the engineers who did produce the first generation of GWR steam locomotive. *North Star*, dating from 1837, was designed by Robert Stephenson & Co for the New Orleans Railway. It was purchased for the GWR at Gooch's suggestion. It was a good engine because its design was not influenced by Brunel's restrictive specification. *Ian Allan Library*

for the disaster, stating that the locomotives were not to his design.

Luckily for the GWR in 1838, it did possess one good engine. This had been designed for a 5ft 6in-gauge railway and ran on the GWR with lengthened axles. The reason for its excellence was that it had a properly proportioned boiler, cylinders and valves. It would have run just as well on axles 4ft 8½in long, and so would all other broad-gauge engines.

The Atmospheric Railway

The 'atmospheric' system of propulsion was not invented by Brunel, nor was he the only engineer to be seduced by its smooth charms. It was suitable for any situation where trains could also have been hauled by a rope winding around a revolving drum. The system worked with a fair amount of success in Ireland, pulling carriages up 1½ miles of single track rising steeply from

The pioneer atmospheric railway in the British Isles was the London & Croydon Railway, when an additional line was laid between West Croydon and New Cross; it lasted until 1847. This 1933 view shows pipes from the L&CR's scheme being dug up and illustrates well the slot (at the top of the pipe) via which propulsion from the piston was transmitted to the train. *British Railways*

Above: The South Devon Railway, engineered by Brunel, was built using the atmospheric principle from the start. It was perceived by railway operators as being beneficial to steam, being quieter and non-polluting at track level. In the event, however, the location of the South Devon line, adjacent to the coast, and other problems with the equipment quickly resulted in its abandonment. On 5 June 1969 a 'Warship' diesel-hydraulic with an engineers' special passes the remains of the atmospheric pumping station at Starcross. *T. W. Nicholls*

Below: One consequence of Brunel's use of the atmospheric principle on the South Devon Railway was that it had to be built at sea level, as the SDR's technology would not permit gradients. Whilst this resulted in the dramatic coastal route at Dawlish, the downside has been an almost constant problem with coastal erosion and the threat of high tides. This view, taken on 23 February 1966, shows work in progress repairing damage; more recently Railtrack has undertaken much more significant work, but, with the threat of rising sea levels, the problems at Dawlish can only get worse. *British Railways*

Kingstown Harbour to the top of the cliff at Dalkey, where the carriages were attached to the main-line train to Dublin. The Stephensons, Daniel Gooch, Brunel and William Cubitt — amongst many other engineers — saw it working, but only Brunel and Cubitt thought it was the perfect way to run a main line. They could not have been more wrong, for the following immensely practical reasons:

1. A train could not be routed from one line to another whilst under power because the tube was in the way of the wheels during the crossing movement.

2. The diameter of the tube and the pressure of the atmosphere dictated the limits of the haulage power of the system. Heavy trains would always run slower than lighter ones; if greater power were required the entire tube would have to be replaced with one of larger diameter and the pumping engines increased in size in order to cope with the increased volume of air to be extracted from the tube.

3. For the cost of one pumping station a great many locomotives could be purchased.

The ultimate consequence of Brunel's choice of the broad gauge was that, some 50 years after the Gauge Commission had ruled in favour of the standard gauge, the GWR was still in the process of converting broad-gauge track to standard gauge, delaying the creation of a truly national network and incurring expense for the GWR. Here track is being converted to standard gauge towards the end of the broad gauge's life in 1892. *Ian Allan library*

4. Shunting under power was impossible; atmospheric trains could not reverse.

Brunel was told all these things by his friends Daniel Gooch and Robert Stephenson, not that he should have needed to be told. In attempting to do what was hopelessly impractical he lost the shareholders of the South Devon Railway nearly £500,000 and saddled the railway forever with some of the steepest, most fuel-expensive main-line gradients in Britain.

Brunel's legacy

Brunel was in many ways his own worst enemy, abusing the contractors and thus the navvies upon whom he relied to build his undoubtedly majestic and strategic railway route. He developed an unfortunate reputation for not paying his contractors. As a result, reputable contractors became wary of working for him, and the railway took longer to build and cost more money than it need have done. Thomas Brassey did not work for him until much later, and the greatest of all contractors up to 1845, William McKenzie, never worked for him at all, even though Brunel spent 18 months looking for a contractor to dig Box Tunnel, the completion of the railway being thus delayed by at least that long. The contractor he finally found, George Burge, was not the best but the most desperate, and more time was lost through his inability to get on with the job — a situation made worse by Brunel's refusing to pay him.

Brunel also refused to pay highly competent building

contractors such as Samuel Peto and Hugh & David McIntosh. Peto had the contract to build the huge Wharncliffe Viaduct, and the McIntoshes that for raising the approach embankments. At one stage Peto was owed £100,000 and had to borrow money to complete the work. When it was completed — on time and within budget — Brunel insisted on going through every invoice to argue and reduce the price; by these means he took a year to pay and knocked £8,000 off the price originally agreed. Peto also had to pay the interest on the money he had borrowed to complete the work, and it appears almost as if Brunel mistreated him simply because he could. When Brunel asked the McIntoshes to take over the work of bankrupt contractor William Ranger, west of Bath, David McIntosh declined, but his blind father, Hugh, was persuaded. Again, Brunel refused to pay for the work, because he believed he was being over-charged (even though the price had been agreed by contract), and

in 1842 the GWR was sued by the McIntoshes for £100,000. Brunel refused to certify payment, and the case dragged on to 1865, six years after his death. By 1863 the GWR was labouring under heavy debts and was in no position to pay McIntosh. On 28 June 1865 the Vice-Chancellor of England, Sir John Stuart, ruled that if the non-payment continued one day longer it would constitute a fraud and ordered that the GWR immediately pay the estate of Hugh & David McIntosh (both of whom had also died by this time) the sum of £100,000, with 23 years' interest and costs. The sum paid out to the McIntoshes was a serious drain on the GWR's funds, and in 1866/7 the railway was on the verge of bankruptcy, paying its shareholders in shares rather than cash. In 1867 its directors asked the Disraeli Government for a loan of £1,000,000, but, unlike today, the Government told the privately owned company to sort out its own affairs. Needless to say, it did.

Brakes

Brunel was concerned to obtain high speed on railways but was unconcerned about providing safe signalling or efficient brakes. With his usual blithe nonchalance when confronted with a subject which did not interest him, he told a Parliamentary Committee that brakes on the Great Western Railway were 'tolerably useless'. With his knowledge of the power of the atmosphere, he could surely have invented a vacuum brake. Daniel Gooch, the best locomotive engineer of his generation in Britain, was unable to make a proper brake. In common with all other locomotive engineers, he believed that to force a brake block against the rim of a locomotive's wheel in motion would cause the revolving axle to twist out of the retarded wheel. Gooch's contribution to solving the problem of brakes was to fit a 'sledge brake' to a couple of 4-4-0 tank engines he designed for the South Devon Railway in 1849. This brake consisted of two 'sledges', placed between the coupled driving wheels. The fireman applied the sledges to the rails by a hand-turned screw. The braking effect was all that might be expected from a sledge, but the jacking effect was great as the two sledges, forced onto the rails, took the weight off the wheels and lifted the engine. The tendency to rip the rails from the longitudinal sleepers as 38 tons of locomotive went tobogganing along their surface can easily be imagined. Furthermore, the friction face of each sledge soon developed a rail-shaped groove. At pointwork this groove, pressed hard to the rails, tried to follow the diverging rails, and there were some derailments in which the engine, wheels eased up from the rails, was dragged sideways by its sledge brake.

British trains remained without proper brakes until at least 1876, and some railways did not so equip themselves until 1889. Yet this was not because a power system was unavailable; it was a question of cost. The companies thought it was cheaper to have the occasional accident than to go to the expense of brakes; in 1865 only 23 passengers were killed of the 252 million who travelled by train. The earliest power brake used on a British railway was Kendall's air brake, installed on the North London Railway in 1850. An air pump attached to an axle on the guard's van pressurised a reservoir to 45psi. A safety valve on the reservoir hissed air continuously as the train sped along. When the brake was required the driver or guard opened a cock and allowed the pressure to enter the train pipe and so to the brake cylinders. No other railway company thought this worthwhile.

In 1868 the 'Irish Mail', worked by the LNWR — the greatest railway company in the world, GWR excepted — was wrecked at Abergele through a lack of any brakes. Among 32 passengers killed were the Duchess of Abercorn, the Marquess of Hamilton and Lord and Lady Hamilton. The most powerful railway company in the world was apparently unconcerned, however, and in 1870 was responsible for one third of all railway crashes in Britain. Its directors, like those of all the railway companies, objected to the expense of fitting brakes, but in 1871 they allowed their new Chief Mechanical Engineer, Francis Webb, to install the most ludicrous and pathetic brake ever to be fitted to a British train — it was not worth the outlay. The Clark chain brake was almost as dangerous as not having a brake at all, the least of its disadvantages being that it applied only one brake block to each wheel of a carriage.

The Clark chain brake was activated by a clutch fixed on the rotating axle of the brake van. The clutch, when released, came into contact with a hitherto stationary shaft. Attached to this shaft was a chain to the brake blocks. The shaft was rotated by the revolving axle and wound up the chain like a windlass, so pulling on the brakes. With the brakes fully applied, the clutch must have been slipping, generating noise and heat. The engine itself had no brakes, but there was a screw-down handbrake on the tender. The brakes on five carriages could be operated from one brake van in which a guard would ride, so trains were made up of sets of five carriages. If a train had 10 carriages the brake van was marshalled at the centre, the van having two trigger levers to operate the brake on the carriages in front and behind. If the train had 15 carriages — not unusual — there had to be another brake van (and guard) at the rear of the train, to operate the brakes on the last five

carriages. The brake on a 15-coach train was thus operated by two guards, bringing the risk of a coupling breaking if the rear guard braked before the front guard. If the front guard acted first, the rear five coaches would run violently against the braking front part. In the case of a breakaway there would be no brake, and the carriages would be out of control.

In 1876, under the auspices of the Board of Trade, trials at Newark of a variety of brakes showed that the Westinghouse compressed-air brake — the invention of an American, George Westinghouse — and the Sanders-Bolitho automatic vacuum brake were the most effective in stopping a train and were fail-safe. If the 'train pipe'

carrying the compressed air or the vacuum were ruptured, the brakes were automatically applied to the whole train. Even after the evidence of the Newark trials, Sir Richard Moon, Chairman of the LNWR, insisted that the chain brake was the finest in the land. An article in *The Engineer* of 5 October 1877 stated that Moon and Webb had 'obstinately refused to act on the suggestions of the Board of Trade, which are not far removed from commands, to install a proper brake'. After the derailment of the Great Northern's 'Flying Scotsman' at Bawtry in 1879, when the brakeless carriages ran amok, Moon was reported as saying: 'We hear a lot about self-acting brakes, but no man in his senses would risk his trains to a

The consequences of a lack of brakes can be dramatic. In October 1988 two Class 31 locomotives — Nos 31202 and 31226 — ran away at Cricklewood yard after vandals had released their brakes. On the morning of 28 October the damage was all too evident as the locomotives blocked the North Circular Road at Staples Corner. *Brian Morrison*

self-acting brake.' In another incident, at Warrington, the track had been taken up and the signals a mile to the rear were at Danger, but the 10am Euston–Edinburgh express, equipped with Clark's chain brake, was unable to stop in a mile and fell off the end of the line.

In 1880 Webb patented an improvement whereby an extra brake block was added, so that the braking was achieved by a 'clasping' action of the blocks. The Clark brake became the Clark & Webb brake and Webb received a royalty for every carriage fitted with the brake which gave him a vested interest in retaining it for as long as possible. The Clark & Webb brake was used on the Caledonian Railway, a close ally of the LNWR.

Even with the Webb improvement it was not up to the job. At Rutherglen in 1881 and at Lockerbie in 1883 the chain brake failed its passengers. After Lockerbie the Caledonian installed the Westinghouse air brake and insisted that the LNWR fit that brake to all West Coast Joint Stock carriages. Only when all carriages on the LNWR had been fitted with his patent version of the chain brake did Webb agree to throw it out and install not the excellent Westinghouse but the near-useless Smith simple vacuum brake, condemned as dangerous after the Newark trials. This then became 'the best brake', according to Moon.

In 1874 Earl de la Warr, on behalf of the railway trade unions, presented to Parliament a Bill which, if enacted, would have made 'block signalling' and automatic (ie fail-safe) brakes obligatory. With 120 railway directors in one or other of the Houses of Parliament, the Bill was defeated, and many more lives were lost, culminating in the great disaster at Armagh on 12 June 1889, when the Smith brake killed 78 and injured 250 out of 600 passengers. As a result the trade unions' Bill of 1874 was revived and put into Parliament on 3 July, receiving the Royal Assent on 30 August as the Regulation of Railways Act. This made it illegal to run passenger trains without a fully automatic, fail-safe brake and gave the companies 18 months to comply. However, even the Act was a blunder, in that it did not enforce power brakes on goods trains, which remained highly dangerous to every other train on the line, and it did not lay down which of the two suitable systems should become standard. Both systems were in use, leading to operating inconvenience. The vacuum brake was good and it was cheap, the air brake much faster in operation but also much more expensive.

The Railway Act 1921, which grouped the railways into four main-line companies, presented an opportunity to standardise on the Westinghouse air brake, but the proportion of vacuum- to air-braked vehicles was 3:1, so it was not done. Under the 1955 Modernisation Plan money was available to fit the air brake to new mineral wagons and thus speed up coal-train movements and increase the capacity of the railway. The Technical Development Committee recommended that this should be done and that the air brake should become the standard on British Railways. The British Transport Commission agreed. Unfortunately, only the Southern Region General Manager was willing to implement the conversion, all the others being so implacably opposed that the BTC backed down — when it actually had the power to insist — and the vacuum brake continued to be fitted to wagons for another 10 years. In 1967 in Britain 1.5 miles was the minimum distance required to bring an vacuum-braked express to a stand from 100mph. In 1936 a German passenger train running at 98mph, equipped with an air brake and not undertaking a special braking demonstration, stopped dead in 0.85 mile.

Burning Coal

Coal was the original fuel for locomotives from 1802 until 1829, during which period locomotives had one large-diameter flue to accommodate the fire and the hot gases to the chimney. The coal gases were slow moving through the wide flue, and the thick coal smoke and air had time to mix and ignite; the smoke nuisance was not intolerable, given that these early engines worked on private railways between a colliery and a river or canal. When, in 1829, Henry Booth suggested the use of a separate 'firebox' with 'a cluster of fire tubes', the small volume of the furnace and the relatively high speed of the coal gases through the tubes prevented the coal smoke from mixing properly with air to cause its combustion, and there arose the serious problem of thick black smoke. Coke — coal with the smoke-producing bitumen already burned off — then became the fuel but was more expensive than coal and had less energy per pound, so more of it had to be burned to achieve the same results as coal.

In hindsight, the complexities in which the early (and not so early) locomotive engineers became involved in order to burn coal in locomotive fireboxes seem quite remarkable. In 1841 a Mr Hall of Nottingham invented a system for forcing air into the firebox. This consisted of continuing the bottom row of 16 boiler tubes to the outside of the locomotive at the smokebox end, where they appeared as a row of bell mouths; had they been trumpets a fine tune could have been played on them. The idea was that the forward movement of the engine forced air, jet-like, into the firebox to help ignite the coal gases. In 1841 the 16 'trumpets' were installed in an engine called *Bee* on the Midland Counties Railway but did not work as well as hoped. A few weeks later, Hall devised an arch of firebrick which extended across the firebox from side to side from the front tube plate, backwards, over the blazing coals — it was just above the air inlets and below the main body of the boiler tubes. Hall's intention was that this 'brick arch' should deflect incoming air from the 16 trumpets — and from the ashpan below the fire — over the surface of the burning coals to dilute the hot gases with oxygen and cause their combustion to produce a clear exhaust. The engine

certainly hummed after that. The chimney emitted a clear exhaust (a great novelty), but the smokebox glowed red-hot because of the high-temperature gases generated by the action of the brick arch in the firebox. This result was so alarming that the brick arch was discarded, but fitting a larger smokebox would have been a better idea.

Locomotive engineers continued to busy themselves with highly complex and expensive fireboxes in an effort to solve the mystery of how to burn relatively inexpensive and highly calorific coal in their engines. The solution was to increase the volume of space available to allow the hot gas and oxygen to mix to effect combustion — to 'consume the smoke'. John Dewrance was first on the trail, in 1845, with a 'double firebox': coal was burned in the rear 'box and the smoke passed through tubes into a second 'box which was the combustion chamber, the theory being that, when the chamber became hot from the furnace in the other compartment, it would cause the gases to ignite, give off their heat and emerge invisibly from the chimney. However, this system did not work well until an iron deflector plate was added to the tube plate (in the same position as Hall's brick arch) to delay the gases so that they had more time to ignite, but the fact that the device worked reasonably only when the arched deflector was added did not register with the engineers — perhaps it was too simple; the enormous complexities of the double fireboxes were much more interesting to engineers.

Since 1853 J. E. McConnell of the LNWR, Joseph Beattie of the LSWR and John Cudworth of the South Eastern Railway had been experimenting with complex and highly expensive firebox/combustion chambers. In McConnell's and Cudworth's designs the partition between the two was longitudinal, giving the fireman two doors through which to throw coal. Beattie's system of 1853 was immensely complex, incorporating two fire-boxes, one in front of the other, and a combustion chamber. The rear firebox burned coal and had a transverse tube (known as the 'water bridge') connecting each side of the firebox; the forward 'box burned coke. There were two fire doors, one above the other, the fireman shovelling coal through the upper door and coke through the lower. As well as reaching the fire through the

ashpan, air was drawn in through hollow firebox stays at each side of the furnace. The smoke from the coal furnace passed through a hole into the coke furnace, where the intense heat ignited the smoke, and all this passed into the combustion chamber, where burning was completed. The fuel efficiency of the Beattie locomotive fitted with this system was excellent, although construction and maintenance costs were high — not to mention the difficulties experienced by the firemen and fire-droppers who had to work with such an awkward system.

Edward Wilson was wrongly credited with inventing the brick arch when he applied it to a new design of his own on the Oxford, Worcester & Wolverhampton Railway in 1858. The Midland Railway engineers took note, experimented, and in 1860 perfected the form and use of the brick arch. The Beattie double firebox remained standard on the LSWR until 1877 — almost 20 years after the brick arch had been proved to be all that was needed to make coal burn efficiently in locomotive fireboxes.

Appointed Locomotive Superintendent of the LSWR in July 1850, Joseph Beattie was one of the engineers who sought a means of burning coal in the firebox without contravening the law requiring all locomotives to consume all their smoke. McConnell, Cudworth and Beattie all sought solutions; typical of Beattie's work was the 'Vesuvius' class of 1870, of which No 17 *Queen* was constructed the following year. *Ian Allan Library*

An almost exact contemporary of the 'Vesuvius' class was the '800' class of 2-4-0 built for the MR, illustrated by No 808 here. In 1859/60 the MR had adopted the cheaper and less complex brick arch and deflector plate, which pattern would ultimately be adopted universally. *Ian Allan Library*

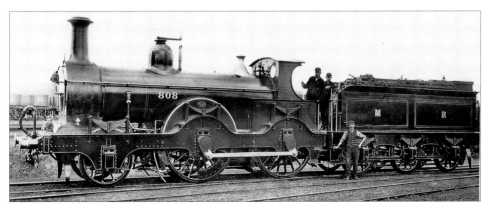

The Tay Bridge

The first bridge over the Firth of Tay was designed for the North British Railway by Thomas Bouch (1822-80) in the years 1869-71. Bouch was born in Thursby, Cumberland, and attended the village school, where his schoolmaster introduced him to the fascination of mechanical engineering. Aged 17 he became an assistant to George Larmer, an assistant engineer to Joseph Locke on the Lancaster & Carlisle Railway. Having worked for Larmer for four years, in 1844 he became a resident engineer under Jon Dixon on the Stockton & Darlington Railway, before becoming Manager and Engineer of the Edinburgh, Perth & Dundee Railway in 1849, in which year he designed the world's first train ferry. This commenced operation across the Firth of Forth in 1850, and another was introduced on the Tay in 1851; though very expensive to operate, they were a great success.

By this time Bouch was well respected and decided to set himself up as a self-employed civil engineer; it was at this point, however, that his ambition for fame and wealth took over, and it could be argued that he was now more concerned with building his reputation than his bridges. Like George Stephenson, he was a practical man, but his formal education in science and mathematics was lacking, and, like Stephenson, he had to delegate much of the detail work to better-educated assistants. Over the next 20 years he was Engineer to 300 miles of railway, including the 50-mile-long South Durham & Lancashire Union Railway linking the Stockton & Darlington with the LNWR at Tebay by way of Stainmore Summit. On this line he is credited with designing the 346yd-long, 196ft-high Belah Viaduct and the 246yd-long, 161ft-high Deepdale Viaduct, although both were designed in detail by Bouch's assistant, an Edinburgh mathematician and engineer by the name of Robert Bow. Into their design Bow incorporated his principles of girder proportions, which then became the accepted standard, but Bouch took the credit, souring the friendship between the two men. The viaducts endured ferocious winds and lasted until 1962, when the line was closed and they were dismantled. But these two viaducts had a better design than the later Tay Bridge: their iron columns were arranged with a broad base, straddling the ground like a Sumo wrestler, those of the Tay Bridge being perfectly upright, with legs close together on a narrow base; the metal in the Belah and Deepdale viaducts also had the advantage of being made from flawless materials.

In 1862 the Edinburgh, Perth & Dundee Railway was absorbed by the North British Railway. For the line to cross the Tay conveniently for Dundee and the East Coast route to Aberdeen, Bouch had to plan for a bridge 10,000ft long

Taken in 1877/8, this view shows well the original single track through the High Girders during the period of the bridge's construction.
A contractor's locomotive can be seen heading towards the photographer.
British Railways

After construction, the original Tay Bridge was subjected to rigorous testing; unfortunately, the testing failed to replicate the conditions that were to pertain in December 1879, when the High Girders were brought crashing down into the chilly waters of the Tay estuary. *British Railways*

— 186yd short of two miles — across the relatively shallow Firth of Tay, whose greatest depth was about 50ft. The bridge was to cross some of the most storm-swept water in Britain, the tidal races in the shallow water being nearly as ferocious as the winds above. It was a terrible place for a bridge, and Bouch's job was to design it to withstand the strains.

Employed by Bouch to carry out trial borings into the riverbed on the line of the proposed viaduct was one Jesse Wylie, who informed Bouch that solid bedrock existed under 20ft of sand for all but 250yd across the Firth. How Wylie could have come to such a conclusion is a mystery, since the bedrock disappeared into the depths about 500yd offshore, but Wylie apparently found what he was paid to look for, and Bouch did not question such a convenient result.

Bouch designed the bridge as a series of 89 spans, varying in length from 27ft to 200ft, supported on brick piers founded deep in the supposed bedrock and rising right up to girder level, to carry a single line of railway. To make the detailed design calculations he employed Allan Stewart, whom he described as 'a better mathematician than myself', but 'according to my directions and orders', being careful to keep his copyright on the job.

Bouch and Stewart completed the design in 1871, whereupon Bouch at once began working out his plan for the crossing of the Firth of Forth and left the supervision of the construction of the Tay Bridge to one Henry Noble — a skilled bricklayer recommended by Joseph Bazalgette, engineer of London's sewer system.

The contract for building the Tay Bridge was awarded to Charles de Bergue on 1 May 1871, and construction work commenced shortly thereafter. After 14 brick piers had been built out into the water it was discovered that the bedrock had disappeared, leaving only a weak, stratified formation of sand and hard-packed gravel or conglomerate. The original brick piers were now too heavy for the riverbed, so, to reduce weight, Bouch redesigned them, terminating the brick portion a few feet above the water line and capping each with four courses of stone. Around the head of each capped pier he arranged cast-iron columns: where the spans to be supported were short, the columns were arranged in groups of four; larger spans were supported by groups of six. These columns, of 12in diameter, were cast in 10ft lengths with end flanges so that they could be bolted together to create the necessary height. They were cross-braced with wrought-iron strips to create, in effect, a tube. The iron bracing strips were bolted to projections which were cast as part of each column; the holes in these lugs were conical — just as they came from the casting — and ought to have been drilled parallel so that bolts could be pushed through with a firm, overall, proper fit. The light columns, rising 80ft or more above the water, were held down to the masonry by massively strong iron bolts, but the bolt heads were embedded beneath only two courses of stone blocks — 1ft of cemented masonry for an 80ft column, with the bridge girders on top of that.

Although the bridge was now to be made largely of iron, Bouch left Noble the bricklayer to supervise the manufacture of its components and their erection as a bridge. The contractors, anxious to make the largest profit from their quoted price, did a rough job, and Noble had to trust their word.

The violent power of the ferocious gales in the Firth of Tay was perfectly well known, but no-one in Britain knew how how much strength to add to a structure to withstand side pressure from a strong wing, and such allowances as were made were guesswork. In 1869 Bouch had made inquiries with the Board of Trade as to what allowance should be made for wind pressure in girders up to 200ft in length. Col Yolland had replied: 'We do not take the force of the wind into account when open lattice girders are used for spans not exceeding 200 feet in length.' Bouch tamely accepted this, as if the Board of Trade were better engineers than he, and made no allowance at all, in the Tay Bridge design, for the force of the wind. (Allan Stewart stated later that he had made an allowance for a wind pressure of 20lb/sq ft; Bouch claimed that he had no idea that this had been done, which makes one wonder how much he knew about a bridge that was supposedly his design.) In 1873, in preparing his designs for a bridge over the Firth of Forth, Bouch asked the Astronomer Royal what allowance he ought to make for wind pressure and was told 10lb/sq ft — a figure endorsed by a committee of five leading bridge engineers. In France and the USA an allowance of 55lb/sq ft was made for wind pressure.

On 2 February 1877 two 245ft-long iron girders, of the 13 known as the 'High Girders', had been lifted to the top of the columns to which they were to be attached. Before they could be bolted down, a gale sprang up, the men were forced to take shelter or be blown away, and both the heavy girders were blown into the sea.

Contractor Charles de Bergue was ill and in financial difficulties when he tendered for the bridge and submitted too low a price, in order to get the job. He was then short of money to build it, the strain on his health increased and he died. His widow and daughter attempted to keep the job going, but Bouch refused to accept women in charge — it having escaped his notice that a man had given him faulty information regarding the bedrock — and the women were forced to give up the contract, which was then awarded to Hopkins Gilkes, of Middlesbrough. This company was founded by one Joseph Dodd, a Member of Parliament whose financial sharp practices had already caused his name to feature in several cases of fraud. Bouch had a large shareholding in this company — but at least they were all men.

The iron used in the castings for the Tay Bridge components was of a type known as Cleveland No 3. Possibly it was cheap. Certainly it was well known for its sluggish nature when molten: it did not run quickly, and this allowed air to become trapped inside the sand mould which was to be filled with the molten iron. The trapped air prevented the iron from occupying all the space in the mould, resulting in reductions in the diameter of the columns and even cavities within. Edgar Gilkes, head of the iron-founding firm, delegated the responsibility of making sound castings to his Chief Engineer, Albert

Grothe, who knew little about iron casting and in turn passed the buck to his foreman at the foundry, Fergus Ferguson. The latter should be immortalised for saying at the subsequent Inquiry: 'I wouldn't say it was the best but it was not what you would call terribly bad iron.' Ferguson had to throw away 200 castings, but where the holes or depressions in the columns were not 'terribly bad' he filled them with beaumontague — a compound of iron filings, beeswax and lampblack — which was then painted over, making the defects invisible. Noble either did not know about this or allowed himself to be fobbed off by Ferguson. Bouch apparently had no idea of the shameful state of the bridge components, because he did not supervise: once the design was complete he lost interest, more or less, in the Tay Bridge because he was far too busy pursuing his greater glory in the Forth Bridge; he left a bricklayer to supervise a company of fraudulent iron founders in which he held shares. Meanwhile Albert Grothe, without the knowledge to supervise iron casting, found himself the job of publicist and gave many talks to convince people that the bridge was safe, asserting that it would require a wind pressure of 90 tons to upset the bridge and that no such wind existed.

In February 1878 the Board of Trade's Inspector, Maj-Gen Hutchinson, Royal Engineers, spent three days examining the bridge, after which he reported that 'the iron work has been well put together both in columns and girders, the lateral movement is slight and the structure shows great stiffness'. He also wrote: 'I should wish if possible to have an opportunity of observing the effects of a high wind when a train is running over the bridge.' In three days of looking, the Major-General had not spotted that there were broken and slack-fitting braces between the groups of columns; he passed the bridge as safe, and it was opened for passenger traffic on 1 June 1878. On 20 June 1879 Queen Victoria passed over the bridge in the Royal Train, and the following week Bouch went to Windsor to be dubbed a knight by Her Majesty in honour of his great work.

Sir Thomas was retained by the NBR at a salary of £105 per annum to supervise bridge maintenance. So how did the great man carry out his duty? He simply passed the buck to Henry Noble and seven unqualified maintenance men; the NBR's civil engineer's responsibility was confined to maintaining the track. The NBR was making large sums of money from traffic over the bridge yet wanted to cut its costs. In September 1878, after an inspection by a diver of the brick piers, Noble agreed to the NBR's request to reduce his costs by reducing his maintenance team to four men; he then took on the role of diver, for which, of course, he had had no training. In January 1879, at the request of the NBR, Noble cut the men's wages by 5s a week, and in June he agreed to sack two more men in September.

When Noble found an iron column rocking on its securing bolts, he personally paid for some iron and had it hammered in under the column to stop it moving. He hammered slivers of metal into bolt-holes to steady loose bracing. Bolts holding the bracings came undone and fell out, the holes they passed through being merely rough apertures in the casting. Vertical cracks appeared in the columns, one of them being 4ft 6in long and $\frac{1}{8}$in wide. Noble probed these with a piece of wire and found them to go deep into the column. He duly reported this to Bouch, who came to inspect on 23 December 1879 and approved. While there, Bouch can hardly have failed to notice the shaky state of the bridge, yet he remained smugly self-satisfied.

Maintenance men and a painter who had worked on the bridge described later how rivets had been falling out and that bracings had fallen off or were loose. John Evans, a temporarily employed painter, testified that the bridge shook so much that his paint pots jumped and fell into the water. He said that he hung a plumb-line off the bridge just to watch its antics as trains crossed; local trains were worst, because they were racing the ferry. The bridge swayed laterally, and each train was preceded by a wave form in the track. The state of the bridge was common knowledge to everyone connected with it and to people living nearby, who could hear the racket as it shook itself to pieces. The bridge had been tugging at its securing bolts for months.

On 28 December 1879 a westerly gale was blowing at Force 11 — 70mph — against the bridge. The bridge could already move; now it was being forced back, pulling on its bracings, pulling against the securing bolts embedded in the brick piers. A train from Edinburgh to Dundee was crossing, and as it entered the High Girders at the south end the ferocity of wind increased suddenly. It was this freak and fantastically powerful burst of wind that forced the bridge over: the defectively cast columns snapped at their weak points; others, exerting 88ft of leverage, uprooted the thin layer of masonry in which their fantastically strong iron bolts were embedded. Inside a span of the High Girders, the entire train fell 88ft into the sea. Everyone on the train —75 men, women and children — perished.

At the Inquiry, Albert Grothe — he who had, month after month, told public meetings that no wind in the

A dramatic view of the broken piers of the High Girders, recorded shortly after the collapse of the bridge. The city of Dundee can be seen on the distant shore. *British Railways*

The calamitous collapse of the High Girders is evident in this view northwestwards, towards the then largely undeveloped area of Dundee along the Perth and Blackness roads. Today the foundations of the original High Girders can still be seen alongside the replacement bridge. *British Railways*

After the collapse of the bridge, the North British Railway initially reverted to the use of ferries across the Tay, but a new bridge would be constructed alongside the old — indeed, partly constructed from material salvaged from the original bridge — and, more than a century after the replacement bridge was opened, still provides an essential link on the East Coast main line. On 16 February 1988 the 14.40 service from Dundee to Edinburgh passes Tay Bridge South 'box. *D. M. May*

At low water it is still possible to see the foundations of the piers for the original bridge, as evinced in this view looking north on 14 July 1961. *M. Pope*

world could blow the bridge down — was asked:

'What occasioned the catastrophe?'

'A strong wind.'

'That was the sole cause?'

Bearing in mind the need for economy at all times, he replied simply:

'That alone.'

A less economical man — a man with some remorse — would have continued: '… acting upon a badly made bridge for whose construction I was largely responsible.'

The Inquiry Report was published in July 1880 and found Sir Thomas Bouch solely responsible for the disaster. He had been stupidly negligent, building glory for himself while depending largely on others' brains. He had been fed false information as to the seabed by Wylie. Britain's

engineering community had had no proper advice to offer him regarding wind resistance and had approved his flawed designs for a bridge over the Firth of Forth. Hopkins Gilkes, the fraudulent iron founders who had deceived him, were not censured but were bankrupted by their own negligence, although they were protected by the law from the full force of their creditors' wrath.

Sir Thomas Bouch never accepted that it was his design and negligence that had caused the failure — even though he must have been aware of the parlous state of the bridge; he blamed the gale for derailing the train and believed it was the train that caused the bolts to be pulled out of the masonry. So much for his engineering sense. Having been unwell for some time, he died a broken man on 30 October 1880.

J. C. Craven

John Chester Craven had had a long career in the locomotive/engineering world before he became Locomotive Superintendent of the London, Brighton & South Coast Railway in December 1847, aged 34. He had been apprenticed to Robert Stephenson in 1829 and passed into the employ of locomotive builders Fenton Murray & Jackson, leaving that firm with high rank in 1837 when a strike broke out. Craven was probably the cause of it; to say that he was a martinet would be an understatement. In an era when harsh, arbitrary and totally unfair behaviour on the part of railway managers — and locomotive managers in particular — was normal, Craven was an absolute swine, violent and deliberately cruel. His grand-daughter wrote: 'My grandfather was entirely devoid of sentiment and if he possessed hidden depths no-one dared to plumb them and he did not realise their existence. His Gods were money and railroads and to these deities he devoted his life's work.' His attitude towards the human race was so ice cold and uncompromisingly awful that one wonders whether he was not mildly insane: he was unable to relate to people — only machines. Asperger's Syndrome, perhaps. Rather than love people he loved his engines, and each one was a treasured individual to be loved and cosseted — until he got tired of it. Not for him the ranks of interchangeable (and thus relatively cheap) spare parts assembled into the form of a locomotive — oh no! He would design an engine for a job — an engine for the Midhurst branch, an

engine for the South London line. The former would be an 0-4-2 with inside frames and the latter an 0-4-4 with outside frames. Then, when he got bored with this or that engine, he would rebuild it, apparently for fun — an outside-cylindered, inside-framed 4-4-0 well-tank engine was rebuilt as a double-framed 2-4-0 side-tank engine.

There is no denying that Craven engines worked well. His finest were the pair he designed in 1862 for working the LBSCR's 'crack' train between London and Brighton. They looked similar to an early Dean Single on the Great Western and were named *London* and *Brighton*. Both lasted for more than 20 years and were very fine, modern engines. However, he never repeated the design and built such a variety of locomotives as has never been seen before or since on any railway — he must have cost the shareholders a lot of money. The nearest Craven came to a standard class of engine was when he designed — and had Brighton Works construct — 17 2-4-0 tender engines with 6ft wheels and outside frames. Beyer-Peacock was given the order to construct 12 more, and the drawings were sent to Manchester. The 12 came home to Brighton, identical in every detail; the 17 constructed at Brighton had all been altered as they were built, and, while they all looked similar, they were all different in detail — there was no interchangeability of parts between any of them.

Craven was a man who could not bear to be contradicted or frustrated in even the most indirect way. His

In October 1866 Brighton Works completed No 230, one of Craven's designs of 0-4-2T for use on the Midhurst branch, although it spent only six months operating on that route before transfer in April 1867 to New Cross for operation in South London. After almost a decade in London it migrated again to the Sussex Coast before being withdrawn from Brighton in 1881. *Ian Allan Library*

John Chester Craven, seated with his wife and two daughters on the right of the picture, with a newly completed 'Crystal Palace' tank forming a backdrop, at Brighton Works in 1858. *Ian Allan Library*

orders had to be obeyed. Fallible humans were a sore trial to him — he could not design people nor rebuild them when the fancy took him. His footplate crews were forced to work very long hours for low wages and usually worked trains when they were exhausted; they were absolutely terrified of him. On 27 November 1851 there was a head-on collision at Ford swing bridge between a moving goods train and a stationary passenger train, the driver of the goods train, Pemberton, having over-run the signal; his engine was badly smashed, but he was thrown clear and lay, knocked about the head, at the foot of the embankment. Pemberton had already had one interview with Mr Craven over some minor incident which had not been his fault. Craven had put real fear into the man on that occasion, even though he had finally exonerated him

Craven was not averse to obtaining additions to his eclectic collection of locomotives from a variety of sources. In late 1867 he was offered a small 2-4-0 built by Kitson & Co and displayed by the company at that year's Paris Exhibition. Designed, apparently, for operation overseas — probably in a warmer clime than that of southeast England (as it was originally provided with a sunshade roof) — the locomotive was available without a tender for £2,000. Acquired by the LBSCR and numbered 248, it subsequently became No 463 *Hove* and was re-boilered and modified under Stroudley's regime (in which condition it is seen here), surviving until 1893. *Ian Allan Library*

from all blame; what would Craven do now that he (Pemberton) had smashed up two of his precious engines? Believing that Craven would prefer it if he were dead, Driver Pemberton took out his pocket knife and attempted to cut his own throat. His guard, Burgess, found him undertaking the operation and wrestled the knife from him — whereupon the driver got up and staggered into the river, fully intending to drown himself. The guard waded out and dragged him back.

Pemberton was duly put on trial for manslaughter. The Judge, however, was scathing regarding the LBSCR's primitive signalling arrangements, and the jury found Pemberton Not Guilty.

During Craven's long tenure of office (1847-69) the Brighton company had more than its fair share of expensive accidents. The trains had no brakes and the signalling was primitive. The Board of Trade recommended installing Lock & Block, but this was not done — the men had to be kept alert, according to Craven and his ilk, who continued to work them 15-18 hours a day and cut their wages when they ran late or over-ran a signal. Many were the additional costs inflicted on the London, Brighton & South Coast Railway by the pride and arrogance of J. C. Craven. He finally relinquished his post in December 1869, and many were the smiles and sighs of relief as this tyrant passed through the works gates for the last time.

F. W. Webb

John Ramsbottom replaced J. E. McConnell as Locomotive Superintendent of the London & North Western Railway. Ramsbottom, arguably, was a nasty piece of work, although his engines were very sound. The measure of his beastliness is demonstrated by the fact that he went to great lengths to destroy McConnell's professional reputation with the LNWR Board by the simple method of altering his engines so that they did not work properly and then blaming McConnell's design for the faults. Ramsbottom's second in command at Crewe Works, from 1856, was the 24-year-old Francis Webb. Webb was such a brilliant engineer that he had been appointed Chief Draughtsman at the age of 22. He was also a hard man, his only loves being work and money. In 1856, with his life before him and Ramsbottom already into middle age, Webb could see himself in his chief's chair in 15 or 20 years. As the years passed he became increasingly impatient with his subordinates, and then, quite suddenly, in 1864, he resigned and went 'outside'. Maybe Ramsbottom had had enough of his pretensions, said 'I'm not dead yet' and shown him the door. In 1871 Webb returned to his old job as Ramsbottom's second-in-command. Three months later Ramsbottom had gone, and Webb moved into his seat with the very modern title of Chief Mechanical Engineer.

Severe and autocratic and a lonely, lifelong bachelor, Webb was also a very great engineer, incredibly energetic and capable. His first passenger engines were based on Ramsbottom principles. They were all similar to look at but came in three groups — in ascending order of power, the 5ft 6½in-driving-wheel 'Precursor', the 6ft-driving-wheel 'Whitworth' and the 6ft 7in 'Precedent'. All were excellent machines, and the 'Precedents' did most of the really hard work on the LNWR from 1874 until 1904.

In 1879 Webb was present at the Institution of Mechanical Engineers when M. Mallett read his great paper on compounding. Webb was bowled over by what he heard. He stated at that meeting that he would build compounds on the three-cylindered layout, with a central high-pressure cylinder and two low-pressure cylinders outside. That, of course, is the opposite of good practice, because the inside cylinder would be charged with high-pressure steam twice per revolution of the wheels but would have to supply steam to the two low-pressure cylinders, equalling four charges of steam per revolution. He must have realised this, for he did not carry this plan into action but did the opposite.

Francis Webb was something of a megalomaniac. He was not stupidly conceited — he was an amazingly clever engineer — but was possessed of a total and absolute belief that whatever he thought must be right. So it was that he became a certifiable compounding freak. Using the exhaust steam from a smaller cylinder in a much larger cylinder to get the last of the heat energy from the steam is a good idea, and some British engineers made it work well. But not Mr Webb. For various difficult technical reasons his system was mediocre, and he made the situation worse by going beyond what was sensible. He it was who introduced into Great Britain, if not the world, the 'divided drive' locomotive — the apparent 2-4-0 which had the leading driving wheels driven by the inside, low-pressure cylinder and the trailing driving wheels by the two outside, high-pressure cylinders, so that the engine was, in fact, a 2-2-2-0, or 'Double Single'. All such compound engines were built to his patents, and he was able to claim a royalty from the LNWR for each one of them. No wonder, therefore, that these dubious machines proliferated.

The first of Webb's Double Singles was introduced in 1882. Named, with great imagination, *Experiment*, it was used on the 'Irish Mail' south of Crewe — a prestige job previously carried out by the excellent 'Precedent' class. When *Experiment* finally managed to start (and *Experiment* always had great difficulty in starting), it surged fore and aft — a motion it imparted to the train, because the 26in-diameter low-pressure cylinder was driving for only half a revolution of the wheels, and for the other half the 11½in-diameter, high-pressure outside cylinders had the job all to themselves. Delays to the 'Irish Mail' became the order of the day. The following year saw the construction of no fewer than 29 production models derived from the badly designed *Experiment*, the first being named with the same heavy logic — *Compound*. They had 13in-diameter high-pressure cylinders but were

'Experiment' class 2-2-2-0 No 315 *Alaska* was built in February 1884 and survived until July 1905. This view shows well the arrangement of the cylinders and driving wheels: the high-pressure external cylinders powered the rear driving wheels, whilst the low-pressure inside cylinders drove the front. *Ian Allan Library*

otherwise a continuation of the *Experiment* design, with a 'Precedent'-class boiler and 6ft 7½in driving wheels. They had the same failings as the original locomotive, including great difficulty in starting from rest.

The 'Compounds' were intended to replace the simple and practical 'Precedents' on the best expresses, but in fact the latter had to be coupled to the front of the 'Compounds' to assist them in their heavier duties to/from Preston and Carlisle. Webb, having got into a hole, decided to dig deeper and designed a larger version of the same thing. These 'Dreadnoughts', the first of which appeared in 1884, had higher boiler pressure and larger cylinders but reduced driving-wheel diameter — 6ft 3in. However, the smaller wheels did not assist starting from rest; the problem here was internal — back-pressure in the small cylinders. They still had to be piloted by a 'Precedent' or a 'Precursor' on the heavier trains, but 40 of these incompetent engines were built.

All engines designed by Francis Webb from 1880 onwards had their fireboxes fitted with his patent 'water bottom' — a typically clever but useless Webb device, like his chain brake. The water bottom was a downward extension of the inner and outer walls and water space of the firebox, and a horizontal floor containing a water

space connected the vertical sides. The space thus enclosed occupied the space conventionally taken up by the ashpan, and ash from the fire fell onto the horizontal 'water bottom'. Webb claimed that this device increased the heating surface of the firebox by 40sq ft, but as the extension was not in contact with the furnace this seems unlikely. In winter these water bottoms picked up spray when the engine passed over water troughs and arrived at their destination encrusted in ice. But the claim that they constituted additional heating was never retracted.

In 1884 Webb began to try out three-cylinder, divided-drive compounding on innocent tank engines. First he seized on a Beyer-Peacock 'Metropolitan' 4-4-0T which had fallen into his hands and rebuilt it to his patent. The second was a new engine, the 'Webb Patent Engine for Suburban Work'. A 2-2-2-2, it was put to work hauling City of London stockbrokers and similar types between Broad Street and Mansion House, but after a week the high-class passengers were in uproar. The engines had the very worst fore-and-aft surging movement of any of the Webb compounds, and, as there were so many stations to stop and start from, the occupants of the carriages were subjected to a thorough shaking at every one. The engines were despatched to where their awful tugging motion did not matter — to the Buxton–Manchester line, there to shake up simple mill operatives. Notwithstanding this ghastly failure, Webb had another engine built to the same design but with small wheels for freight work. He exhibited this at the Manchester Exhibition of 1887, presumably as an example of LNWR excellence — yet all

four tank engines were so bad that not even Webb could justify them, and as soon as their boilers had worn out he ordered them to be scrapped.

In 1889 Webb introduced the first of 10 'Teutonic'-class compounds — effectively 'Dreadnoughts' with 7ft 1in driving wheels. These were followed in 1890 by a yet larger edition of the same theme — 10 'Greater Britain' 2-2-2-2 compounds. Although they were bigger engines, they did not perform as well as the 'Dreadnoughts' — and they were none too bright. In 1895 Webb began a final series of the 'Greater Britain' design, running on 6ft 3in driving wheels. These were known as the 'John Hicks' class and were a complete flop; very soon relegated to stopping trains, they were scrapped within a few years.

During the 1888 'Race to the North', wherein the 10am Euston–Edinburgh competed with the 10am King's Cross–Edinburgh, the 'Compounds' were not used at all. On 6 August the schedule to Edinburgh was reduced to 8 hours and the engine that took the train out of Euston was a Ramsbottom 'Lady of the Lake' Single, *Waverley*, dating from 1862. The train was early into Crewe and was taken on by one of Webb's thrashable 'Precedents' of 1874. In the 1895 'Race to the North' — which took place each night from 15 July to 22 August — the 'Dreadnoughts' worked the racing 8pm Euston train from Euston to Crewe, quite often taking a 'Precedent' or even an old Ramsbottom 'Lady of the Lake' for a pilot. North of Crewe the train was rarely entrusted to a Compound. On 28 occasions it was hauled by a 'Precedent', with a 'Whitworth' or another 'Precursor' for pilot. There were, however, six nights when a 'Teutonic' Compound took the train from Crewe to Carlisle. On 13 July *Teutonic* itself

No 1301 *Teutonic* was the first of the 'Teutonic' class and emerged in March 1889. These locomotives had larger boilers than the earlier 'Experiment' class and larger-diameter wheels than the 'Dreadnought' class. The 10 locomotives of this class were withdrawn between 1905 and 1907. *Ian Allan Library*

took a light train of 7$\frac{1}{2}$ coaches[1] and was slower by 8min than a 'Precedent' with 12$\frac{1}{2}$ coaches. On 16 August the 'Teutonic' *Coptic* took 13 coaches from Crewe to Carlisle in 173min, yet on 2 August the 'Precedent' *Mercury* had taken virtually the same load in 3min less. On three other occasions when a 'Teutonic' was used north of Crewe, its load was six coaches, and on a fourth occasion 4$\frac{1}{2}$ coaches, for which the journey time from Crewe to Carlisle was 147$\frac{1}{2}$min.

In 1893 Webb brought out the first of a fleet of three-cylinder 0-8-0 compound freight engines, following these in 1895 with a triple-expansion locomotive, *Triplex*. This had a high-pressure cylinder on the right-hand side, the next, larger cylinder between the frames and the largest cylinder, to take the lowest pressure, on the left. The engine found it difficult to start even without a train, so Webb commandeered it to haul his Inspection Saloon a couple of times a year, but it was not scrapped until George Whale took over from Webb in 1903.

Webb spent most of his career trying to get rid of coupling rods, although the same lifetime's experience should have taught him that the 'coupled' engines he

[1] This refers to the weight of the train rather than to the actual number of coaches, a standard coach being deemed to weigh 20 tons.

had designed — even though they were smaller — were as good as (or better than) the larger, uncoupled compounds. In 1897 he exhumed Richard Melling's 1837 idea of a friction roller as a means of coupling the drive on a locomotive. To this end he removed the coupling rods from one of his loyal and trustworthy 2-4-0 engines and fitted Melling's roller between the two driving wheels. How he could have thought this was a good idea is a secret he took with him to the grave. Anyhow, the idea was a flop, the time and money expended on it wasted and Melling's roller was returned to the dustbin of history.

After this dramatic failure Webb lost his faith not only in uncoupled driving wheels but in the whole idea of three-cylinder compounds — a conclusion it had taken him 15 years and hundreds of engines to reach. He now began to design and build four-cylinder, compound, coupled passenger and freight engines. Between 1897 and 1901, 40 'Jubilee'-class four-cylinder compound 4-4-0s were turned out from Crewe Works. However, despite all his years of experience, Webb made their cylinders too small and they had to be replaced; boiler pressure similarly was too low and had to be raised. These improvements were incorporated in the 'Alfred the Great' class, of which 40 were built between 1901 and 1903, but still the valve-gear was wrong and had to be replaced by Webb's successor, George Whale. Whale tried to make Webb's 4-4-0s work, but all 80 had been scrapped by the end of 1908. Also in 1901, 40 four-cylinder compound 0-8-0 freight engines had been built. They were front-heavy, with a huge four-cylinder-in-line casting, and some were later given a pony truck by Whale, but after a few years they were either scrapped or rebuilt to conform to Whale's standard, simple-expansion, two-cylinder 0-8-0 'G' class (the 'Super Ds'), which was so successful that it survived into Nationalisation.

It was clear as early as 1885 that the Webb version of compounding did not work. The design was so flawed that it was frequently the case that, with the engine at rest on its train, only one high-pressure piston was receiving steam; the steam ports on the low-pressure cylinder were then closed by the valve and the other high-pressure cylinder had its connecting rod on a front or back dead-centre. With only one cylinder taking steam, the train could not be moved. It was not at all unknown for a team of men with a lever under the wheel — a 'pinch bar' — to lever the engine forward until the wheels turned and the valves moved to a better position.

There was more. The inside, low-pressure cylinder was a single engine, acting on the leading crank axle. Because the wheels were not coupled there was no momentum to assist the single-cylinder in turning the crankshaft past dead-centre. If the engine were started by the movement of the inside cylinder, that cylinder would be foiled as the crank came onto dead-centre and the train would stop again. (Richard Trevithick's single-cylinder engine of 1803 had a flywheel to bring the crankshaft over dead-centre.) It was possible for the high-pressure cylinders to start the trailing driven wheels and make them spin or slip while the leading pair were motionless; it was also possible for the leading and trailing driving wheels to spin in opposite directions! Webb was only too aware of this, but the knowledge could not have been painful to him because he never cured it by coupling the wheels and thereby — maybe — turning the engines into something half-decent.

Because Webb was the complete dictator of Crewe Works and town, absolutely nobody dared make a suggestion, much less actually criticise him, yet he cost the LNWR's shareholders tens of thousands of pounds by constructing ridiculous engines. More than 200 expensive, complicated and not terribly useful express and freight engines were turned out steadily year after year from Crewe for 15 years because no-one — not even the Board of Directors — dared contradict Francis Webb. George Whale began scrapping all the three-cylinder Webb compounds as soon as he took office in 1903, and the last of the passenger engines had been melted down by 1907; the freight compounds were rebuilt as 'G' class 0-8-0s. The four-cylinder compounds Whale tried to put right as described, but it was a waste of money, and in 1908 he began rebuilding or scrapping both classes.

But just to be fair: in his will Webb left £196,158 net — the proceeds, to a great extent, of the royalties he received for his mediocre inventions — almost all of which he bequeathed to charitable causes, including £10,000 to found a hospital for the poor and needy and £75,000 to build and endow an orphanage for the children of Crewe.

A Clash of
Personalities

Alexander Newlands was a Highland Scot, born at Elgin in 1870 and educated in the town. Leaving school at 15 or 16, he was articled to a civil-engineering firm in Inverness. After serving his time, in 1892 he joined the Civil Engineers' department of the Highland Railway at Inverness. He immediately proved himself to be a highly capable man, and in 1893 he was appointed Resident Engineer for the HR's extension from Dingwall to the Kyle of Lochalsh, where he also oversaw construction of a deep-water harbour and pier. He got the job completed in four years and was at once re-directed to the widening of the Highland main line north of Blair Atholl. This he completed in 1899, when he was promoted to Chief Assistant Civil Engineer.

Frederick Smith was a Geordie, two years younger than Newlands. He had served an apprenticeship in the Gateshead works of the North Eastern Railway and came to the Highland Railway in 1903 as Works Manager at Inverness. Smith was a very advanced thinker in loco-motive science — O. S. Nock ranks him fourth in the history of British locomotive engineers, after Churchward, Bowen-Cooke and Gresley — but, like Churchward, he was somewhat lacking in the social graces, took easy dislike to his fellow officers and was well and truly rude to them whenever he felt irritated by them. One man in particular who irritated Smith was Alexander Newlands, Chief Civil Engineer of the Highland Railway and thus responsible for the track, bridges and structures. There occurred between Smith and Newlands what O. S. Nock describes enigmatically as 'a rather tactless incident' concerning a new turntable for Inverness shed. Smith was the Geordie outsider, a clever mathematician and thinker; Newlands was the insider, born and bred in the district, a conservative individual who did not understand , or trust — Smith and his fancy mathematics. After the turntable incident, the two men were not on speaking terms, which is a bizarre way for intelligent men and chief officers to run a railway.

Never a rich railway, in 1915 the Highland was in a very bad way because it now had to cope with greatly increased wartime traffic at a fixed income from the Government. Much more powerful locomotives were required to cope with wartime loads, and engines from other companies were sent on loan. F. G. Smith thus designed a very powerful 4-6-0, to be known as the 'River' class, which would have been so heavy as to put a maximum of $17^3/4$ tons on its driving axles — a greater load than was permitted by Newlands, out of respect for the under-line bridges. The HR 'Castle' and 'Clan' 4-6-0s had axle loadings of $15^1/4$ tons, the highest permitted by the Civil Engineer, but if the effect of the 'hammer blow' — the additional weight brought on the track once per revolution of an engine's wheels, caused by unbalanced masses in the motion — is taken into account, those engines brought a weight of almost $21^1/2$ tons to bear onto the arches of under-bridges.

In his 'River' design Smith had, with great math-ematical skill, so carefully adjusted the balancing of the wheels that their hammer blow was two thirds that produced by a 'Castle' and almost half that of a 'Clan', so that the total weight put on the track by a 'River' class was almost exactly that by a 'Castle' and a full ton less than by a 'Clan'. Smith, knowing that his new design was in fact within weight limits, did not bother to inform Newlands of its weight. Of course, the very high figure of deadweight on its axles soon leaked out of the Locomotive Drawing Office and across to the Civil Engineer's office. Newlands, burning with a desire to punish Smith's contempt for him, decided (a) to say nothing until the first engines arrived, and (b) to then ban the new engines on the grounds of their being too heavy for the line and thus drop Smith in the ordure with the directors. As the first two engines arrived from the builders, Newlands was waiting and condemned them on the spot. A furious row erupted. Smith told Newlands and the Board that the engines were within limits due to the reduced hammer-blow, but Newlands did not know about the hammer-blow effect, did not understand the mathematics and refused to be persuaded. The Highland Board, egged on by Newlands (who was not, at that time, Chief Engineer) gave Smith a week in which to resign or be dismissed. He resigned and returned to Newcastle, where he established himself as a consulting engineer.

The Highland Railway thus lost the best engines and

Above: An early illustration of a 'River', in grey for photographic purposes, recording the first phase of the class's life. The photograph's original caption epitomises the position: 'Built for HR; found to be too heavy and sold to CR'. In one short sentence, Frederick Smith's design was condemned. *Ian Allan Library*

Below: All six locomotives of the class passed to the Caledonian Railway (as Nos 938-43) and in 1923 to the LMS (14756-61). Ironically, post-Grouping (and after very limited upgrading to the route) they were allocated to Perth shed for operation over the ex-HR main line to Inverness. This view shows CR No 941 (later LMS No 14759), which should have become HR No 73 when constructed in 1915; it would remain in service until February 1939. The last of the class was withdrawn in 1946. *Ian Allan Library*

the best locomotive engineer it ever had. The directors were pleased because they thought they had saved their railway from untold damage and because they sold the two engines and the four still under construction to the Caledonian for £500 each more than they paid for them. Meanwhile the Highland Railway 'Clan' and 'Castle' classes went on their thundering way, thumping the track with a combined dead-weight and hammer-blow which exceeded that of Smith's design.

In 1923 the Highland Railway became part of the LMS, and in 1927 Newlands was appointed to succeed the great Louis Trench as Chief Civil Engineer. In 1923, meanwhile, a National Bridge Stress Committee had been established which included Newlands. An essential part of the committee's work was to study the effect of hammer-blow on bridges, and the work of F. G. Smith on the better balancing of locomotives was the textbook used — with the Highland 'River' class as the example. The correctness of Smith's calculations was proved beyond doubt, Newlands' nose was well and truly rubbed in the ordure and the six 'River' class were transferred *en masse* from the Caledonian section to the Highland as soon as the study was complete.

In 1932 a paper was read to the Institution of Civil Engineers on 'Impact in Railway Bridges' by Prof Inglis OBE, MA, LlD, FRS, MICE. This was a very mathematical essay on the effect of the hammer-blow on bridges. Alexander Newlands was deputed to open the discussion on the paper and chose to do so with what might be regarded as polite scorn for the complex mathematics: 'I feel rather nonplussed by being asked to open the discussion because I am unworthy to sit down at the mathematical feast which the Author has spread. It would have been useful to have had the abstruse mathematical deductions reduced to simple formulas as could be used in drawing offices.' Newlands went on to deny that the hammer-blow was 'of large significance' in damage to the track. He was followed by three eminent engineers, who disagreed with him and also said that the Professor's formulae for calculating the extra tonnage entered on the track by the hammer-blow were 'handy', 'useful' and 'manageable'. Sir Henry Fowler, Chief Mechanical Engineer of the LMS, came as close to the bone as politeness would allow when he stood up and said: 'Locomotive engineers have known for many years of the existence of hammer-blow but bridge engineers have not worried about it until recently. I have had frequently to argue [with the Civil Engineer] about a few hundredweight of static load per axle whilst I have never had hammer-blows up to 15 tons questioned.' Newlands had banned a whole class of fine engines and lost a man his job over 2 tons' deadweight on an axle whilst remaining oblivious to the 7-ton hammer-blow some lighter engines were giving his track six times per second.

A Line Too Far . . .

The Invergarry & Fort Augustus represented a line to serve the Great Glen in Scotland and, despite the lack of population, was fought over by both the North British and Highland railways as both sought to safeguard their territory. Promoted by an independent company, the line reached Fort Augustus before the money ran out and the company was forced to appeal to other companies to take over the route. Initially, from its opening in 1903, the line was operated by the HR, but the latter soon pulled out, leaving the line to be operated by the NBR from 1907 until Grouping (services being suspended between 1911 and 1913). Passenger services were withdrawn in 1933, and the line closed completely in 1947. This view shows Fort Augustus Pier station — the line's ultimate terminus — in 1905. *Ian Allan Library*

Of the various eccentric lines built in the British Isles, few were more peculiar than the Listowel & Ballybunion monorail constructed in Ireland. This was the invention of a Frenchman, Charles Lartigue, and many miles of this were laid down in France and the French Empire — at Tours, Algiers and Tunis — and also in Argentina. This view features a Kitson-built locomotive at Ballybunion station and shows how the locomotives and rolling stock straddled the track — an arrangement of which the late Heath Robinson would have been proud. *Ian Allan Library*

The Victorian era was replete with engineers and entrepreneurs who were larger than life. One such figure was Sir Edward Watkin, who developed a grandiose scheme for converting the Manchester, Sheffield & Lincolnshire Railway into an international enterprise courtesy of a new main line to London, a link with the South Eastern & Chatham Railway and a Channel Tunnel to produce a network linking Britain and France. The last 'great' main line to be constructed in Britain, the Great Central was nicknamed 'Gone Completely' as a result of its impecunious existence after its opening in 1892/3.
Here a southbound DMU crosses over the West Coast main line at Rugby with a service to Rugby on 17 October 1964. Ironically, the closure of the line in the 1960s as a through route can, in hindsight, also be seen as a blunder (see page 93). *P. H. Wells*

The North Eastern Railway monopolised the movement of coal from the Yorkshire coalfield to the port of Hull, and, as capitalism abhors a monopoly, mine owners sought to get a better deal through the promotion of a competing line — the Hull & Barnsley Railway. Once characterised as terminating in the spoil heaps of Cudworth, the line was incorporated in 1880 and opened five years later. Right from the start there were financial problems, and merger with either the Midland or North Eastern was mooted, although it was not until 1 April 1922 that the company's independent existence ceased, with the NER taking over.
With the inevitable rationalisation of duplicate lines, the H&BR was closed progressively, this being completed by the late 1960s. Here, on 2 September 1957, 'WD' 2-8-0 No 90623 heads a down coal train at Little Weighton. *D. R. Smith*

The Paget Engine

It would be difficult for any Chief Engineer to have as an assistant a young man of quite volcanic intelligence and forceful personality, and when that young man is also the highly precocious son of the Chairman of the Company, with his eyes on his Chief's job, the difficulty could well become very painful. In 1903 Cecil Paget, the 29-year-old son of the Chairman of the Midland Railway, was Assistant to R. M. Deeley, Works Manager at Derby. Deeley was not only 20 years older than Paget; he had been in the service of the Midland Railway since 1875 and was entitled to consider himself Johnson's successor, but Johnson had let it be known that he wanted Paget as his replacement when he retired. In September 1903 *The Locomotive* had even announced that Paget had been appointed. The atmosphere between Paget and Deeley was therefore somewhat tense and would not have been improved when they were thrown into each other's company for several months when they were sent away to the United States to see what good ideas they could glean from US

railway works and locomotive practice. In January 1904 S. W. Johnson retired and was replaced by Deeley, while Paget became Works Manager at Derby, with Henry Fowler as his assistant.

Cecil Paget had a tempestuously lively brain, which in the years 1904-6 conceived the most revolutionary steam locomotive that Britain was to see for 40 years. This machine was a 2-6-2 with outside frames and coupling rods and eight inside cylinders. It had a very wide firebox with two doors to enable the fireman to reach the back corners with his shovel. The interior front, side and back walls of the firebox were lined with firebrick; there was no water space around the firebox, but the firebox crown was covered by water. The driving wheels were of 5ft 4in diameter. Boiler pressure was 180lb. The eight cylinders were horizontally opposed, 18in x 12in stroke, and were cast in two blocks of four. Like those in a motor car, the cylinders were open at one end. The pistons within the cylinders were of the 'trunk' variety — the type normally used in diesel or petrol engines: the connecting rod came up inside the piston head and was attached to it by a 'small end' and 'gudgeon pin'. The pistons moved horizontally fore and aft longitudinally. The two opposed cylinders on the left were offset to each other but

Typical of the designs produced by Richard Mountford Deeley for the MR during his short tenure was No 990, a 4-4-0 compound constructed in 1909. *Ian Allan Library*

Paget's experimental locomotive, No 2299, emerged in 1908 but failed on one of its initial test runs and would never run again. It was to spend the next seven years gathering dust before being scrapped in 1915. One factor in the locomotive's construction — and presumably a cause also of Deeley's resentment of it — was the fact that Paget had been promoted from Works Manager at Derby to General Superintendent of the Midland, thereby becoming boss of his erstwhile superior. *Ian Allan Library*

connected to form a single chamber; the two on the right were the same. Steam was distributed through a cast-iron rotary valve turning in a bronze sleeve. Steam entered between the respective piston heads and forced them apart — they moved in contrary directions. The leading block was placed between the leading and middle axle, the trailing block between the middle and trailing axle. The middle axle was thus driven by four pistons, and the leading and trailing axles each by two pistons. The concept was brilliant: the reciprocating parts were in perfect balance, so there would be no hammer-blow to the track. The engine weighed $74\frac{1}{2}$ tons, with about $18\frac{1}{2}$ tons on each driven axle. It had a big, well-heated boiler and looked likely to be very fast and smooth-riding. Its Achilles heel would be the sleeve valves, because they were made of dissimilar metals which would expand at differing rates when heated; very careful lubrication would be necessary.

The Midland Railway Chairman— Paget's dad — agreed that Derby Works would build the engine but at the private expense of Cecil Paget. However, it proved more expensive than Cecil had estimated, and, when his own funds ran out, the work was completed at the Midland's expense.

In 1907, while his amazing engine was still under construction, Cecil Paget was promoted out of Deeley's office by the General Manager, Guy Granet, to be General Superintendent of the entire Midland Railway. Granet could see that Paget's mind was one that could visualise complex problems and then conceive ways of solving them. Amongst other matters, he expected Paget to find a way to prevent the appalling delays to goods trains on the heavily overcrowded Midland trunk lines. To this end Cecil Paget brought into his office a very fine District Traffic Inspector, J. H. Follows, and set about organising an intricate and comprehensive Central Traffic Control System, the first of its kind. There was to be no fragmentation of responsibility — everything to do with the running of trains was under one hand — so Paget was also given control of the day-to-day running of the locomotive fleet and the crews, which had always been the province of the Locomotive Superintendent. Deeley's job was thus seriously diminished — he had commanded 18,100 men but now he was in charge only of the 5,000 at Derby Works. His erstwhile assistant was now in command of his engines, footplatemen and running sheds, which must have caused steam to issue from his Victorian ears and made him even more jealous of 'golden boy' Paget.

Mr Deeley was a man of great scientific achievement and with no less ambition to create magnificent loco-motives — and, of course, he had been in the business of locomotive engineering for far longer than had Paget. He had begun his working life in 1873 as a pupil-engineer at the Hydraulic Engineering Co in Chester and after two years had transferred his pupilage to the office of S. W. Johnson, Locomotive Superintendent of the Mid-land Railway. Having gained repute as an experimental scientist, he was soon appointed Chief of the Testing Department at Derby. He continued to be associated with research and development work, which involved electrical and hydraulic engineering and chemistry, and he and the Midland Railway's chemist developed the company's water-softening apparatus. Deeley also studied the problems of lubrication, invented an ingenious apparatus to measure the 'oiliness' of oil and wrote a text book on 'Lubrication and Lubricants', which was still a standard work for steam-locomotive engineers in 1945. He was made Works Manager in 1902. While Deeley was responsible for building the Paget engine — which was gaining such nicknames as 'Paget's Folly' and 'Paget's Mistake' — Deeley was busy designing a rival 'wonder-engine'. Remarkably this had a 2-4-4-2 wheel arrangement. His machine was to be a compound tank engine with a boiler pressure of 220lb, a fire grate of 28.4sq ft and four outside cylinders, the two high-pressure cylinders driving the four rear driving wheels and the low-pressure cylinders the four at the front. Despite the 2-4-4-2 description it was not an articulated engine, all eight driving wheels being coupled together.

Paget's engine, numbered 2299, emerged from Derby Works in 1908. Deeley, however, was madly jealous of it and had no intention of doing anything to assist its introduction into revenue-earning traffic. With Deeley — Britain's leading authority on steam-locomotive lubrication — in charge of the trials, the Paget engine's sleeve valves seized solid while the engine was running at 82mph on a test train, whereupon Deeley gladly ordered it off the line, and it never ran again. Shortly afterwards, Deeley's proposals for the compound tank engine were rejected by the Board of Directors. White-faced with fury when he learned the news, he went to the works, collected a carpenter and the man's tools and marched to his office, where he ordered the workman to remove the brass plate bearing his name and title from the door and to place it in his tool bag. Deeley then stalked out of the works and out of the Midland Railway forever.

The Paget engine, which might easily have been an important development in steam locomotion, was destroyed by jealousy. Ignored and forgotten, it remained hidden away under heavy tarpaulins until scrapped for the war effort in 1915.

The Railways Plundered, 1940-5

In April 1937 the Government and the railway companies began negotiations concerning the Government's rent for their railways in the event of a war. The companies were concerned that their fares and charges would be held down when prices rose, as had happened during World War 1. In 1939 the Government gave an assurance that the railway companies could raise their charges in line with inflation, but, despite this, the railways were specifically excluded from the same year's Prices of Goods Act, which permitted all other businesses to raise their prices in line with inflation. On 1 September 1939, under the Emergency Powers (Defence) Act, the Ministry of War Transport took control of almost all railways in Britain.

On 7 February 1940 the Control Agreement based on the previous three years of negotiations was published. The Agreement contravened the 1854 Act, which was still in force and forbade railways to give undue preference to anyone; it also breached the 1921 Act guaranteeing the railways a 'standard income'. The railways would carry Government traffic at a 33% discount, and the total earnings of the companies from civilian and Government traffic would be pooled and shared, the LMS receiving 34%, the LNER 23%, and the GWR and SR 16% each. If the total receipts of the railways exceeded £43.5 million, the balance was to be shared between the railways and the Government; any income above £68.5 million would accrue wholly to the Government!

In the February 1940 Agreement the Government permitted the railway companies to increase their charges to cover costs, which were bound to rise in wartime. Charges had risen 10% by April 1940. In September 1940, however, the War Cabinet broke the Agreement, telling the Chancellor of the Exchequer to (a) reduce the railways' permitted annual revenue to £40 million and (b) prevent any further increase in railway fares and charges. This was to help keep down inflation — a Government imperative — although the part railway charges played in inflation was minimal.

In October 1940 the Chancellor's office warned the War Cabinet: 'It is of the first importance that the railways be maintained in a healthy economic state. The railways are entitled to a modest reward. We recommend that charges be increased and the [original] Control Agreement be preserved.' Instructed to follow orders and cut the railways' income, the Chancellor again warned the War Cabinet: 'Static [railway] charges risk bankruptcy. Higher charges do not give increased profit but reimburse the companies for money actually paid out.' However, having issued these warnings, the Chancellor, without any consultation with the companies, broke the negotiated 1940 Agreement and subjected them to a reduced income. No government had ever put a penny into the railways in their entire history — but now this Government was going to confiscate a large part of their income whilst forcing them to carry a considerable amount of traffic free of charge! No other British industry was subjected to such treatment.

Road hauliers were treated with great respect during the war. Road haulage companies were not brought under Government control and did not have their takings seized by Government, yet — in this dire national emergency — the Road Haulage Association haggled, and it took from January 1941 until March 1942 to agree a price for the hire of 2,500 lorries with which the Ministry of War Transport wanted to form a pool. Meanwhile, the railways were being forced to carry food for nothing! The charges of the uncontrolled road haulage industry rose throughout the war — their charges were not, apparently, inflationary. The thousands of companies took advantage of the wartime demand situation and had to be asked by Government to keep their greed within 'reasonable levels'. The magazine *Modern Transport* noted in February 1940 that road hauliers' charges in this emergency 'were out of all proportion to their increased costs'.

When, in February 1941, the Government needed a pool of thermally insulated containers for the carriage of imported meat, the railways provided 6,317 insulated vans and 2,354 insulated containers for which they were paid nothing because the Government had the use of the railway for a flat rate. The road haulage industry provided 1,241 road lorries which were hired commercially and their owners paid in the usual way.

When American forces began to arrive in Britain from 1942, the work of the railways increased, and in the months leading up to D-Day railway work became ever more intense. The Government increased its income from the increased traffic because of the Control 'Agreement'. Year by year the railways were worn down with over-work, lack of maintenance and a shortage of money. Only absolutely essential maintenance and renewals were undertaken, owing to a shortage of material, lack of labour and the fact that railway workshops were turning out munitions of war. By November 1943 the Ministry of War Transport reported that the railways were dealing with 50% more traffic than in September 1939, with fewer locomotives and men, and all for an annual flat rate rent.

The Government of 1939-1945 and all subsequent governments found it very useful to have control of rail transport. Holding down railway charges allowed a government to give a subsidy to the travelling public and industry. That same industry could — and did — then raise its charges when supplying materials to the railway. This policy was good for votes but disastrous for the railway. Between 1939 and 1945 railway charges rose by 16% whilst the cost of running the railway increased by at least 70%.

Government figures show that the combined revenue earned by the four railway companies between 1940 and 1946 totalled £487,600,000, of which the Government had helped itself to £195,100,000, and the balance was then divided between the four companies. Each railway company received less, after all its work, than the Government — which had done nothing apart from oppressing the companies, grinding them down and leaving them ever more vulnerable to the road haulage industry. There was a real danger of railway bankruptcy, and in 1946 the Government permitted a 33% increase in fares — which still left railway fares well below the rate of inflation — but for which 'extortionate rise' the railway companies incurred the wrath of the newspapers and the population at large.

Bulleid Locomotives

Oliver Vaughan Snell Bulleid was a brilliant if unconventional locomotive engineer who, after 14 years as Assistant to Sir Nigel Gresley, replaced the Swindon-trained R. E. L. Maunsell as Chief Mechanical Engineer of the Southern Railway in 1937.

In 1941 the Southern Railway's Operating Department was desperately short of modern motive power. The SR had put its scarce resources into electrification, and much of its steam fleet was antiquated and puny. On the Western section — the old LSWR network — no fewer than 104 1896-designed 0-4-4 'M7' tank engines were heavily relied upon for goods and local passenger trains on the non-electrified lines and also for hauling empty coaches in and out of Waterloo; rakes of 10-15 carriages

The first of the new 'Q1' class 0-6-0s, No C1, recorded when new, on 26 March 1942. The locomotive weighed 29 tons 6cwt and had a tractive effort of 30,000lb; unfortunately, one of the primary uses for the type — unfitted freights — could prove problematic, as brake power was not sufficient. *Southern Railway*

had to be moved from Clapham sidings or even as far out as Brookwood. Hopelessly overloaded for the latter work and not much better on the other services, the gallant 'M7s' were preventing improvements to the timetable, and a powerful replacement was required.

During 1941 Bulleid designed two remarkable classes of locomotive — the 'Q1' 0-6-0 goods engine and the elegant, streamlined, three-cylinder 'Merchant Navy' Pacific.

The 'Q1' was a powerful shunting, goods and local passenger engine, ideal for replacing an old 'M7', but it had some bad faults. Drivers visibility from the cab was poor when the engine was running tender-first — as a goods engine often did — and, if there was not plenty of coal and water to hold the tender down, when running tender-first at 40mph or so it would jump about in a very scary manner. The Southern had been very sensitive about wobbly engines since a big 'River' class tank engine had wobbled right off the track at Sevenoaks in August 1927; this had been due to imperfections in the track which caused the water in its side tanks to slosh to and fro

through the tanks, setting up a rolling motion which eventually derailed it. The 'Charlies' had another scary problem. In those days most goods trains had no brakes, and the weight of the train was far greater than the weight of the engine. The 'Q1' brakes were not as good as they might have been, and drivers had to be very cautious or they could find themselves heading towards a red light with a very nasty taste in the mouth. This being wartime, such defects had to be overcome by driver skill.

For the 'Merchant Navy' class Bulleid designed a boiler second to none in its phenomenal steam-raising capabilities. Each of the three cylinders had its own valve gear, all three sets of which were placed within the frames; this contrasted with other three-cylinder designs,

With the additional space afforded by the UIC loading gauge, Continental railways have been quick to exploit the potential capacity gains offered by double-deck rolling stock. The fact that the Southern's loading gauge was much smaller did not deter Bulleid in the late 1940s when it came to the construction of a new generation of EMUs. Two units, Nos 4001/2, appeared in 1949 with a double-deck configuration, allowing a capacity of 1,104 — a 33% increase over that of conventional stock. The fact that the passengers were akin to sardines in a tin may have been one reason the design was not perpetuated (although, with the constraints of the 21st-century railway at least one Train Operating Company seems to be dusting off Bulleid's blueprints). *Ian Allan Library*

in which only the central cylinder's valve gear was located under the boiler. To compress the equipment into the confined space available, Bulleid designed a chain-driven variation of the Walschaerts gear. The three sets of valve gear and the central big end were enclosed within a sheet-steel oil bath. The locomotives, being of the 4-6-2 Pacific wheel arrangement, were naturally more prone to wheelslip because of weight distribution than were 4-6-0s, but the 'Merchant Navys' were the most wheelspin-prone engines in the land. The oil bath could not be kept oil-tight, and oil leaked out onto the driving wheels, making wheelslip a certainty. Violent wheelspin damaged the chain-driven valve gear and put valve timing out, to the detriment of performance. The chains which drove the valve gear from the central crank axle sometimes broke, and the heavy, flailing chain would then smash cylinder covers and bend connecting rods.

The middle connecting rod's big end was somewhat delicate and although running in a flood of lubricating oil was prone to failure. Because the big end was enclosed within the oil bath, drivers could not hear the warning knocks of a loose bearing, and the first they knew of trouble was when the big end broke and the connecting rod punched a hole through the oil-bath casing.

Each engine was fitted with a steam-operated reversing gear whereby the driver could adjust the valve travel and so admit more or less steam to the cylinders at the touch

One of the major design faults with Bulleid's designs of Pacific for the Southern Railway was that the smoke-deflectors failed to live up to their name, and the footplate crew would regularly find themselves shrouded in steam. Evincing the problem on 13 January 1952 is unrebuilt 'West Country' No 34102 *Lapford*, seen at Petts Wood Junction with the 9.30am boat train from Victoria. The problem would ultimately be reduced as members of the class were rebuilt. *J. G. Click*

of a small handle. The Southern Railway possessed a very good steam reverser, designed in the 19th century by James Stirling for the South Eastern Railway. Bulleid, however, designed his own, which was rescued from total failure only by the skill of the drivers who had to use it; indeed, it was so temperamental that drivers set it on starting and tried not to touch it again thereafter. In his book *60 Years in Steam* (David & Charles 1986) D. W. Harvey, Shedmaster at Norwich, states that he rode on these engines with 230lb in the boiler and 80lb in the steam chest, because the driver was frightened to alter the reverser. Altering the setting whilst on the run was to risk putting the engine into full forward gear, or even reverse — it was not uncommon for the operating handle to be in the reverse position when the engine was actually running forwards at 80mph.

No 35005 *Canadian Pacific* spent two years on trial at the Rugby stationary test plant as engineers tried to measure its performance. Because the steam reversing gear would not hold the valve travel steady — and

because, in any case, the percentage of 'cut off' indicated on the reverser's scale bore no relation to reality — no measurement of performance could be made, and no report was issued after two years' work. What the tests did make clear was how to turn these eccentric locomotives into truly practical machines.

The engines realised their full, considerable potential only when Bulleid's chain-driven Walschaerts valve gear was replaced with the conventional form of the same gear and his eccentric steam reverse removed in favour of a conventional hand-operated screw. Upon rebuilding, all lost their 'air smoothed' outer casing.

When Bullied first arrived on the Southern his assistant was James Clayton, who had come from the drawing office of the Midland Railway at Derby in 1914 to be Personal Assistant to the SECR's Chief Mechanical Engineer, R. E. L. Maunsell. Whilst at Derby, Clayton had been involved with the construction and testing of the Paget engine and had made his own set of drawings of the locomotive; having retired from the Southern in 1943 he broke the universal silence surrounding this engine, describing it in detail in an article for *The Engineer*. It was then that Bulleid decided to combine his own innovations on the 'Merchant Navy' Pacifics with Paget's to produce what he thought would be the perfect steam locomotive. This was to be the 'Leader'.

Bulleid had something of Brunel about him — he was exceptionally clever, had enormous enthusiasm for revolutionary mechanical ideas and was possessed of great personal charm, which he rolled out to mesmerise those above him into agreeing to his outrageous plans. The comparison does not stop there; Brunel was warned against the atmospheric railway by his friend Daniel Gooch, and Bulleid was warned about his great plan by his Chief Draughtsman, C. S. Cocks.

Asking (in 1944) 'What sort of locomotive may we expect to see in the future?', Bulleid proceeded to answer himself:

'It should be able to run over the majority of the company's lines, it should be capable of working up to 90mph, it should have its whole weight available for braking and adhesion, it should be equally suitable for running in either direction without turning and with unobstructed look-out, be ready for service with minimum of servicing and be almost continuously available, run not less than 100,000 miles between general overhaul with little or no attention on the depot, cause minimum wear and tear to the track and use substantially less fuel and water per drawbar horsepower developed.'

Between 1944 and 1947 Bulleid planned a locomotive which used steam generated by a coal fire and ran on wheels but which thereafter bore no similarity to anything then extant. A powerful locomotive of trustworthily conventional design was required as quickly as possible to ease the Southern's shortage of short-haul motive power. In that era of austerity, when steel for building locomotives and coal for burning in them was in short supply, what better than a tank engine carrying the powerful Bulleid boiler from the 'Q1' 0-6-0? War-ravaged, rationed 1945 was definitely not the time to re-invent the steam locomotive — especially one incorporating items from a failed experiment from 30 years before.

Having made a strategic blunder, Bulleid now made the practical blunder of designing his locomotive of the future a coal-burning, hand-fired machine. He changed the original concept in detail over the two years of drawing-board development, during which time oil-firing was being installed on GWR engines, but he did not even consider a mechanical stoker for this futuristic beast. Sleeve valves on steam locomotives are particularly unnecessary; conventional piston valves would have been a sensible idea, but convention was being deliberately avoided. The exception was the continued reliance on a human fireman, yet a mechanical stoker would have increased the power of the locomotive..

Instead of designing, building and putting into traffic in a matter of months desperately needed new motive power, Bulleid wasted time and money on two years of trials — and errors — and not only with 'Leader'. Other engines were fitted with various bits of the experimental gear, most importantly Atlantic No 2039, which was a good engine taken out of traffic and put through the works to receive new cylinders and sleeve valves so that experiments could take place to see how well the valves worked. Meanwhile the original source of complaint — the 'M7' tanks — continued to struggle and impede the traffic in attempting to perform tasks far in excess of their design capacity.

Construction of the first 'Leader' began in July 1947. Design was slow because Bulleid and his design team kept devising extra 'improvements'; like Brunel, Bulleid was bubbling with bright ideas which led him to change plans that had already been finalised. The main frames were not completed until May 1948, by which time the Southern Railway had become British Railways Southern Region. The cost of the wonderful new engine, even in batch production, would be double what a modern, conventional engine of the same power output would cost, and the project became known as 'The Bleeder', on account of the resources it was bleeding from the railway.

The first and only 'Leader' engine to steam, No 36001, left Brighton Works in late June 1949. It was not the same engine whose design had been approved by the Southern Railway directors in 1946; that design had had superficial similarity to one of Bulleid's Pacifics, being carried on two groups of six driving wheels, and the boiler would have been where one expects to find a boiler, placed along the centre-line of the frames. But the engine that left the works in 1949 had a diesel-type cab at each end, and the boiler was now offset from the centre line of the locomotive so as to provide a corridor from front to back cabs and to the fireman's compartment at the centre of the machine, $2\frac{1}{2}$ tons of pig-iron being added to the 'light' side to balance the machine.

Not only was the engine to be hand-fired; the fireman was expected to do this from a completely closed-in compartment at the centre of the vehicle. Imprisoned therein, he had difficulty in knowing where the train had got to and thus found it difficult to know how to control his firing. Bulleid designed a curved coal-chute from the bunker which sometimes caused the black stuff to jam up. The Southern Region guessed that 'Chinese laundry'

conditions would apply, and, rather than roster (order) a fireman to work there, invited volunteers. Firemen Sam Talbot and Ted Forder stepped forward. Ted Forder always maintained that the engine was no more uncomfortable to fire than was a Bulleid Pacific on a summer's day. Perhaps his nickname was 'Asbestos Ted'. Temperatures of 122°F were routinely recorded, together with a steam-laden humidity. Ted Forder also recalled that the engine's acceleration was better than that of an electric train and its ability to raise a great deal of steam in a short space of time was better than that of a 'Terrier' tank, although, as the 'Terriers' were almost the smallest locomotives in the country, the comparison seems somewhat incongruous.

Ted was an exceptionally loyal railwayman. He and Sam Talbot were firemen of Olympic capabilities, able to shift huge quantities of coal in a very short space of time — for very little reward in terms of locomotive performance. Would every fireman be as loyal?

Not only did the 'Leader' use coal mines of coal for little result; it also used 400 gallons of water per mile because of steam wastage from its sleeve valves — a phenomenon well known to Bulleid from the trials on No 2039 — and burned more lubricating oil per mile than a 'Western' diesel used for fuel over the same distance.

While on trial on 9 October 1949 No 36001, hauling 150 tons and fired by Ted Forder, ran out of steam — even though Ted had the fire so hot that the cast-iron blocks acting as firebricks had melted and the side wall of the firebox was red-hot. The reader will recall that this firebox had water only over its 'crown' or roof, as on the Paget engine, there being no space at the sides of the 'box for

water to absorb the heat and be converted into steam. Conventional firebricks duly replaced the cast-iron blocks. The engine was frequently short of steam, even with three coaches.

On 25 October 1949 a sleeve valve disintegrated.

On 29 June 1950 a crank axle broke.

On 28 September No 36001 took 325 tons from Eastleigh to Woking. With 195psi entering the cylinders and with that supply of steam being cut off after 25% of each stroke of the piston, the engine took 12min to reach 50mph, which seems decidedly pedestrian, considering that the gradient was no steeper than 1 in 252. On the return trip from Woking the engine took $12^{3}/_{4}$min to reach 50mph. The official report states: 'There was no difficulty in maintaining adequate boiler pressure.' Fireman Clayton might have disagreed — he shovelled 1,008lb of coal in that time, averaging 79lb a minute!

On 17 October 1950 the 'Leader' took out a test run with 13 coaches and with a Railway Executive observer — Roland Bond, R. A. Riddles' right-hand man. The engine performed normally and without incident until it was brought to a stand a little short of the water column at Basingstoke. It then refused to move forward or backwards until, after a 6min delay, it decided to stop messing about and responded to its driver's commands. This refusal to start was a common occurrence, and, should it happen when it was on its own, the crew could use pinch bars to move it along the track and get it 'off centres'. It was ironic that it could be moved with ease with a lever between the rail and the wheel — because it was fitted with roller bearings — but it sometimes could not be moved with high-pressure steam, because all cranks were on centres.

No 36001's last test run took place on 2 November 1950, when it took 15 coaches (480 tons) from Eastleigh to Woking. It went steadily up the long bank from Eastleigh, exerting 1,100hp on the drawbar. So what? After four years of effort, a locomotive had been developed which was less effective than a little 'U'-class 2-6-0 which was used to provide a benchmark against which performance could be compared: the 'Leader' burned almost twice as much coal and evaporated 12% less water, overall efficiency of the 'Leader' being 2.82% against 4.72% for Maunsell's very conventional, essentially Churchward-based locomotive.

Up to November 1950 No 36001 had cost £53,210 — some of which had been borne by the Southern Railway — and, even if all its faults had been corrected so that it could be put into service, it would have been less effective than a 'U'. Besides No 36001 there were, in various stages of construction at Eastleigh, four more of these monstrosities — Nos 36002-5; between them these had cost £131,653 and were still not ready to run; their sleeve valves could not be made steam-tight and wasted steam; their fire grates were now too small because of the extra thickness of firebrick necessary in the firebox; they were 20 tons heavier than originally designed, seriously restricting their route availability; their axleboxes were too rigid, placing huge strains on the crankshaft, causing it to break, and also making the engine too rigid when negotiating curves, damaging the track. The off-centre boiler, meanwhile, was detrimental to even weight distribution, correction of which would have required a major redesign, and it was suspected that the welded parts of No 36001's boiler were cracked. Finally, it was extremely unlikely that the SR had enough 'Asbestos Teds' willing and able to fire them. Riddles had absolutely no choice when, on 20 November 1950, he formally recommended that BR's losses be cut and the engines scrapped, whereupon certain national newspapers soundly castigated socialist, bureaucratic British Railways for suffocating brilliant, innovative, private-enterprise ideas.

Following the creation of British Railways in 1948, Bulleid migrated across the Irish Sea to become Chief Mechanical Engineer of Irish transport authority Coras Iompair Éireann (CIE). One of the problems that CIE faced was that Ireland lacked natural resources, and virtually all its coal and oil had to be imported. There was an abundance of peat, but this was already recognised as having a much lower calorific value than coal and creating much more ash and was thus widely perceived as having little practical worth in terms of powering transport. CIE had effectively decided against the electrification of the network — the population density making it unviable — and was therefore looking ultimately to replace its steam locomotives with diesel power. Bulleid, however, decided to experiment with a version of his 'Leader' design to burn peat. In August 1957 emerged No CC1, an 0-6-6-0T similar to but smaller than the 'Leader'. Whilst the locomotive proved both quiet and more efficient than earlier peat-powered locomotives, it was living on borrowed time: Bullied retired the following year, and CIE progressed with its policy of dieselisation. No CC1 was never formally taken into CIE stock and was withdrawn in 1965. The locomotive is seen here at Inchicore alongside more conventional steam-locomotive designs.
J. G. Click / National Railway Museum

Oil-firing

In 1946 the Great Western Railway fitted Laidlaw-Drew oil-burners in the fireboxes of 18 coal-burning freight locomotives working in South Wales. Storage tanks were installed, as were pumps to convey the fuel to the locomotives' tanks. Costs were incurred. The trials were highly successful, and plans were made to extend the trial to express passenger locomotives working in Cornwall and also between Gloucester and Paddington. These trials too were successful — on Sapperton Bank an oil-fired '53xx' was as good as a 'Castle'. Then the Ministry of Fuel & Power heard about these successful but tentative trials and ordered the company to convert 1,217 locomotives immediately so as to save coal, the GWR being expected to find the costs. The loyal and long-suffering company had converted 93 locomotives and constructed oil-fuel depots at some engine sheds when the Treasury heard of the plan. That august institution saw not greater locomotive efficiency and fuel saving but a dollar-exchange problem: oil must always be purchased in US dollars, and Britain did not have sufficient dollars — so it said — to purchase the necessary oil on the scale of the Government's plan. The excellent GWR scheme thus had to be abandoned, and doubtless the GWR bore the costs of the experiment. Had the GWR's engineers been allowed to develop oil-firing unhindered, it seems very likely that the dollar supply would not have been a problem, and as time passed the situation would have changed, allowing more oil-firing to be installed. Steam engines, cheap in construction cost, would have had their maintenance costs reduced, their hours of availability for traffic increased, their haulage capacity increased and thus their running costs reduced. Steam engines could have remained as a cheap and reliable form of traction until replaced, in an orderly manner, by electric traction.

The first GWR passenger locomotive to be fitted with oil-burning equipment was 'Hall' class No 5955 *Garth Hall*, seen here in July 1946 shortly after the equipment was fitted. The oil tank in the tender is clearly evident. The experiment was, perhaps, symptomatic of the debates post-Nationalisation between the Treasury and other Government departments that were in many respects to bedevil the industry: despite the obvious technical attractions of oil-fired locomotives the use of imported oil (as opposed to domestically produced coal), it was swiftly cancelled, and within a couple of years all the oil-burning locomotives were converted back to coal. *Ian Allan Library*

Nationalisation

Nationalisation of the 'Big Four' railway companies was inevitable following the Labour Party's landslide election victory in 1945, since that party had been committed to nationalisation since its inception. The socialistic method of central direction, of pooling resources and of co-operation that had enabled an ever more worn-out railway to continue its vital task was seen to be the system for postwar reconstruction. What Clement Attlee and his Labour Government did not want to know was that Winston Churchill's wartime coalition Government, of which they had been members, had adopted a policy towards the railways which was bound to wreck them. Railways needed a huge injection of capital to repair the damage inflicted by the wartime Government's abuse — and, of course, the damage inflicted by enemy action, although it is arguable that Churchill's Government inflicted more long-term damage on Britain's railways than did the Luftwaffe.

Could the railways have survived as private companies, given the dreadful straits in which the war and the Government had left them? Between 1938 and 1948 railway assets lost about £440 million of their value (at 1948 values), due mainly to wartime neglect. At 1946 prices, wartime damage from air raids on the railway would cost £30 million to repair, and arrears of maintenance would cost £151 million. Then would come the cost of replacing worn-out locomotives and stock that had been kept at work because of the war, to which can be added the modernising of stations. The companies had been unable to attract enough investment before the war, when they were in good order; for postwar investors there were quicker ways of getting a greater profit from a smaller investment than giving their money to the exhausted railway companies.

In the years 1945-7 the railways were in a more-or-less run-down state — the GWR and SR less, the LMS and LNER more — and in no position to attract large-scale investment. In 1946 £100-worth of Ordinary Shares in the LMS and LNER were valued on the market at £29 and £7 respectively; even £100 of the lordly Great Western's Ordinaries were selling for a mere £51. Any new finance for the railways would have to be raised with a promise of guaranteed interest — and at a rate approximating to what other industries were paying. The question is: could the railways have raised the money on the open market? If they had remained in private hands, the Government would have had to repay to the companies their confiscated renewals fund and compensate them with at least a percentage of the £1 billion-worth of free services the railways had rendered. In the immediate postwar years the national economy was in no state to do this, so, as in World War 1, the railway companies had subsidised the entire country. Under the Railways Act 1921 they had been 'Grouped' into four and awarded highly inadequate compensation by a cash-strapped Conservative Government; in 1948 an equally impecunious postwar Labour Government had done the only thing it could — even without its party dogma — and nationalised them.

The Transport Act 1947 came into effect on 1 January 1948, establishing the British Transport Commission (BTC) and below that a series of Executives to supervise each branch of transport, the Railway Executive controlling British Railways (BR). The railway was divided into Regions, but these were under strict orders from the Railway Executive. The Act laid down that: 'All the businesses carried on by the BTC will form one undertaking'. This meant that income from the railways — which generated 70% of the BTC's income — went into the general coffers of the BTC and was used, in part, to subsidise canals, docks and harbours. Once again, the railways were being milked for the benefit of others.

The shares and bonds issued by the railway companies were extinguished by the Transport Act, shareholders being compensated with British Transport Bonds, which paid an annual interest of 3%. This was good news for LMS and LNER Ordinary Shareholders — who had not received a dividend for years — but bad news for those holding 4%-6% guaranteed interest stocks in the GWR, SR, LMS and LNER, who lost not only their annual dividend but also the market value of their holding; in 1946 £100 of GWR 5% Mortgage stock had been worth £142 7s 6d. The total of railway capital to be compensated by the British Transport Commission (BTC) was £1,693,000,000, as stated in the 1947 Act. The great

The end is nigh: on 30 December 1947 a member of staff pastes the poster announcing the 1947 Transport Act in Manchester. The Act had the effect of nationalising 60 railway undertakings, the London Passenger Transport Board, 18 canal and inland waterway companies, ports, harbours, steamers, hotels and more than 20,000 long-distance lorries. *Ian Allan Library*

blunder of the Act was that the railway — the BTC — was obliged to find from its revenue, annually, 3% of £1,693,000,000 and in addition to repay annually to the Government a portion of that capital until the capital sum was entirely paid back. British Railways was obliged by law to buy itself and pay dividends to shareholders who, in many cases, had never received a dividend under private enterprise. On top of this, the law forbade BR to compete with coastal shipping and subjected all BR's proposed price rises to a Rates Tribunal which was supposedly independent but which was in fact an arm of Government. As ever, Government wanted to depress railway charges for political reasons. During the war, the Government had profited out of the railway to the tune of £1 billion — at wartime prices — and had also confiscated £147 million, which formed the combined railways' deferred maintenance and renewals fund. The Government kept for other purposes all that railway

money which it could have used to create a fund to pay the railway shareholders their 3%.

A large part of the road haulage industry was nationalised under the 1947 Act and all road goods vehicles, except those used by traders for their own businesses, were restricted to working within a 25-mile radius of their home base. The British Transport Commission was obliged by the Act to buy the old road lorries from their owners at the cost of a new lorry.

The law also obliged the BTC to purchase from their private owners 544,000 railway wagons, usually wooden, most of them antiquated. The compensation paid to the owners of these museum pieces was the price of a new wagon in 1939. The cost to the BTC for buying scrap wagons at new prices was £43 million. Depreciation was suspended! Surprisingly, for a socialist government, all this was generously biased towards the private owner — in contrast to the treatment meted out to the railway itself. It was as if the old companies had created such a powerful image in the minds of the Labour legislators that they thought railways were all-powerful and could survive any amount of abuse.

Section 4(1) of the Transport Act 1947 includes the famous obligation which has been repeated in subsequent Acts and which has been so often misquoted down the years that it is given here in full:

'The Commission shall so conduct that undertaking and levy such tolls, fares, rates and charges as to secure that the revenue of the Commission is not less than sufficient for making provision for the meeting of charges properly chargeable to revenue, taking one year with another'.

This is wonderfully vague for a law but implies that it would be lawful for the BTC to make a loss in one year if this were offset by profit in another, the BTC being obliged to run the railway efficiently and cover its costs — which, of course, included buying itself from the previous owners.

To make ends meet, the BTC would have to close down loss-making branch lines — just as the private companies would have been obliged to do — and increase its charges up to the rate of inflation, at the very least. But then the Government put major obstacles in the way — line closures were to be subject to long consultations, during which the losses mounted, while increases in passenger fares and freight charges were permissible only with the consent of the Transport Tribunal. In 1949 Lord Hurcomb, the civil servant who had had much to do with Government's abuse of the railways during the war and who was now appointed by Government to be Chairman of the BTC, suggested to railway officers that they reduce the price of tickets to 1¼d per mile — a rate last charged in 1914, when the cost of living was 20% that of 1949. There thus arose the quirky situation whereby the top man was suggesting a course of action illegal under the terms of the Act incorporating the organisation he led.

Lack of income prevented the BTC from paying wages equal to the industrial norm, which led to a shortage of manpower as potential recruits to the railway went to work for the new industries like motor-car manufacture. British Railways was also kept at the back of the queue for the scarce supplies of steel and wood; the motor-car industry was, of course, at the front, due to the need to 'export or die'. The railway tracks were thus maintained to a minimum standard for years after the war, rather than the arrears of maintenance being overtaken. The repair and replacement of locomotives and rolling stock was curtailed. Old rolling stock, of small capacity and frequently under repair, had to be kept expensively in use. And, of course, the public and successive governments blamed BR management for the lack of progress in modernising the railway. That the BTC managed to cover its costs in its first seven years was due to the sensible shackling of the road haulage industry in the role of collection and delivery for railways.

Nationalisation could have worked — and pigs may one day fly. If the Labour Government had treated the BTC with justice, if the Labour Government had properly followed through after passing the 1947 Act and had really looked after the railways as befitted its main choice of transport for the nation, if the Labour Government had not been thrown out in 1951 … But no government had ever treated the railways with justice. The Labour Government believed that dogma was enough and, having nationalised the railways, withheld funds in the general purse and let the railways sink or swim. It was a sorry business.

The 'Britannias'

In 1948 R. A. Riddles, Chief Mechanical Engineer of British Railways, who had begun his engineering career on the LNWR, stated that he would 'buy the form of locomotive that would give the greatest tractive effort for the least amount of pounds sterling'. This meant building steam engines. His enthusiasm was laudable, but he did get rather carried away, designing 12 different 'Standard' types. This itself could be described as a blunder, while there were design blunders in the details, especially in the most prestigious of the Standard classes — the Class 7 'Britannia' Pacifics and the solitary Class 8 locomotive, No 71000 *Duke of Gloucester*. These design faults were not such as would affect day-to-day

performance in the 'Britannias', which were powerful engines, but in No 71000 the faults most certainly did affect performance — so much so that it was considered to be a disappointing failure. The small 'Britannias' ('72xxx'), otherwise known as the 'Clan' class, were not entirely satisfactory and indeed should not have been built since they was only marginally more powerful than the Standard Class 5s ('73xxx'). This design performed well but not as well as the LMS Class 5 on which it was based — a great blunder was not to reproduce the LMS Class 5 and save the cost of two new designs. Below the '73xxx', in diminishing order of size, were the '75xxx', '76xxx', '77xxx', '78xxx', of which probably only one class was needed, and the tank engines from the big '80xxx' to the little '82xxx' and the tiny '84xxx', of which, again, only one was needed. It has to be said that the final BR Standard type, the Class 9F ('92xxx') 2-10-0, was one of the finest designs in British locomotive history, the blunder here being that construction continued until 1960, with the result that some examples were barely run-in by the time withdrawal commenced.

The first BR Standard locomotive, Class 7P6F Pacific No 70000 *Britannia*, emerged from Crewe Works on

2 January 1951 and was quickly followed by 24 more of the class. They were designed at Derby rather than Crewe or Swindon, which was unfortunate for details in the design — although, of course, the general principles of the engine and boiler were descended from Swindon via Crewe.

The 'Britannias' were originally fitted with a rubber-damped drawbar between engine and tender. The usual fore-and-aft surge inherent in any two-cylinder engine on starting and during powerful acceleration was amplified — hilariously so in the restaurant car, with spilt soup and coffee — by the 'squidgy' nature of the rubbery drawbar. This caused so much trouble that Derby was forced to replace its design with that employed by the GWR for 60 years and exported to Crewe — a massive iron drawbar, pin and sprung buffers between engine and tender.

Locomotives of the 1951 batch had fluted connecting and coupling rods (not used by the GWR since the 1890s, because the 'I' section is not as strong as a regular rectangular section) and ran through Timken tapered roller bearings. The axles were of 9in diameter but were hollow, a 4in-diameter hole being bored right through. The GWR had first used this idea in 1922, to improve the tempering of the steel — the weight saving was merely a bonus. But according to the designers of the 'Britannias' the hole was there purely to save weight, which calls into question whether the axles had been properly heat-treated and tempered. The wheels had been pressed onto the axles with a pressure of between 10 and 13 tons to overcome an 'interference fit' between the circumference of the axle and the hole in the wheel. The external diameter of the axle was greater at the roller bearing than where it fitted into the wheel; thus wall thickness was brought close to the limits of safety inside the wheel and the difference in thickness created a weak point in the axle, although the Derby's designers did not recognise this. With the wheel on the axle, a tight-fitting steel peg was driven into a slot aligned in the axle and wheel, using a very heavy hammer, to make it impossible for the wheel to move around the axle. Despite this, in March 1951 No 70014 *Iron Duke*, working a Southern Region express on the Dover run, suffered bent coupling rods when all its coupled wheels shifted on their axles. Six more 'Britannias' suffered this fate in the next few weeks and early in May No 70004 *William Shakespeare,* travelling at high speed with the 'Golden Arrow', suffered a broken coupling rod for the same reason. Only at this eighth, potentially fatal failure did Riddles take the precaution of withdrawing the entire class — 25 engines at this stage — from traffic.

Riddles' cure was to force a steel plug into the axle as far as the axlebox and to replace the fluted coupling rods with the stronger rectangular-section rods. The use of the Timken roller bearing axle box produced a very reliable, almost maintenance free bearing but introduced a point of weakness in the axle. In the later batches of the class, Riddles dispensed with the roller bearings in favour of conventional axle boxes and was thus able to have axles of uniform diameter throughout their length. Some 'Britannias' never experienced wheel shift even when the engineers set out deliberately to try to move them; others of the class were prone to it. The plugging of the axles worked well for three years of very hard work, but then the problem came back to haunt some members of the class. The shift was a matter of a few thousandths of an inch, but this was enough to interfere with the correct running of the engine and require works attention.

The 'Britannias' were originally fitted with a patent device called a Steam Drier. This device had been sold to BR by the Superheater Company and consisted of spiral vanes fitted around the steam-collecting pipe. The idea was that, as steam rose from the boiler water surface, the spiral vanes caused the steam to swirl round vigorously so as to fling water droplets back into the boiler water; thus 'dried', the steam entered the steam-collecting pipe. In practice, however, the spiral plates formed a kind of ladder up which the water climbed before entering the steam-collector. The height of the dome was raised, allowing the steam-collecting pipe within to be raised further away from the water surface, but the trouble persisted.

The Steam Drier was eventually discarded, but even then the problem of priming — water passing from the boiler into the cylinders — did not go away entirely. Water trapped between an advancing piston head and the cylinder cover is as incompressible as a block of steel and causes the same damage. Cast-iron cylinder covers were blasted off at speed, followed by the piston on several engines. Priming also has a very damaging effect on superheater elements, which become clogged with limescale from the water, causing them to split open or burn off at the ends, while the water going on into the valve chests and cylinders destroys the lubrication of the engine. On the Eastern Region, 'Britannia' valve and piston rings lasted only 10,000-12,000 miles, providing constant work for maintenance staff.

Another design fault which beset the 'Britannias' was the use of riveting to attach the crosshead slide bars to the frames. The design was similar to that used by the LNER on its Pacifics (but perhaps they were not attached with rivets), and, compared to the method adopted by the GWR at Swindon and exported to various other railway works, the Doncaster method looked flimsy. Starting from rest with a 400-ton train, the powerful reciprocating thrust of the piston forced the crosshead upwards with a force of 3.8 tons until the rivets holding the slide bars to the engine frames began to work loose. The slide bars rose and fell, causing the piston rod to flex where it entered the crosshead and piston. In 1952 No 70001 *Lord Hurcomb* broke its piston rod while travelling at high speed near Forncett, throwing its left-hand piston through the

cylinder cover. After this a careful check was made on all 'Britannia' piston rods, and several were found to be defective. The rivets holding the slide bar bracket were replaced with bolts, turned to be a perfect fit in the holes through the frames. This solved the problem for four years, but then the brackets started to work loose again.

The 'Britannias' were very powerful, but they were clearly not powerfully designed, and even under the most careful maintenance regimes they managed to do themselves serious, potentially fatal injury every once in a while. The slide bars were badly designed in principle and in detail, because fitters found it very difficult to get spanners on the nuts holding top and bottom bars together. A special thin spanner was issued for this, but thin spanners are not the best for doing nuts up tight. The nuts were not 'crenellated', and the cotter driven in below the tightened nut could not actually prevent the nut from slacking off. The bolts had lugs on them to prevent them turning, but the lugs sometime broke off. The crowning folly was to put the slide-bar bolts in head downwards. On 20 January 1960 No 70052 was on a Glasgow–Leeds sleeping-car train on the Settle–Carlisle route when the slides bar bolts began to fall out, allowing the piston rod to bend. Finally it broke, the piston rod, crosshead and

The 'Britannias' were not the only type of BR Standard to suffer problems. The unique No 71000 *Duke of Gloucester*, with its Caprotti valve gear, was never a success whilst in BR service. On withdrawal, the valve gear was removed for display in the Science Museum and the rest of the locomotive sold for scrap — to Woodham Bros at Barry — where it was to remain until secured for preservation after many years in the open. Long considered to be an impossible scheme, restoration of the locomotive was ultimately completed in 1986. However, the preservationists altered the design of the locomotive's blastpipe and ashpan, such that its performance back on the main line far surpassed anything it achieved when new. *Ian Allan Library*

connecting rod fell onto the ground, was driven forwards and the engine 'pole-vaulted' over it and then continued to drag the wreckage along the track. The crosshead caught in the sleepers of the opposite line and pulled the track out of shape, derailing a passing goods train which then side-swiped the express, killing 5 people and injuring eight.

It is worth noting that the engines designed for the heaviest duties (Classes 70xxx to 75xxx) had the LNER slidebar/crosshead arrangement, while all the others, employed on lighter work, had the very strong and far more sensible GWR arrangement.

The Transport Act 1953

If I may be forgiven for starting with a digression. An observation on the way that history is deflected. In 1950 the Labour Party, led by Clement Attlee, was re-elected with a majority of 17 seats, having received 46.1% of the votes cast. Attlee was soon faced with (a) a serious health problem and (b) a serious situation in his cabinet, with Aneurin Bevan and Harold Wilson in rebellion over the National Health Service. Attlee went into hospital for an operation and came out in November, whereupon he immediately set about bringing the mutineers to heel by calling a General Election. His close ally, Hugh Gaitskell, who might have persuaded him otherwise, was abroad. The election took place early in 1951 and the Labour Party received 48.8% of the votes cast, 0.8% more than were cast for the Conservatives. The Conservative Party was returned to power by a majority of 26 seats. This was a political blunder brought on by fragile health but might also be considered a failure of the British electoral system. It certainly changed the course of British transport history.

With the return of Winston Churchill and the Conservative Party came an apparent upturn in the fortunes of the road lobby. Railways were bound to be distasteful to the Conservatives because railways tend towards the socialistic: quick profits are no part of railway operation. They are collective, communal and work best when centrally controlled. They oblige people to gather at set times, throw them all together and take over their persons for the duration of the journey. The new Government was bound to be antipathetic to railways for the reasons stated above and as can be seen from internal government correspondence concerning the development of roads.

The Minister of Transport, Alan Lennox-Boyd, believed the 1947 Act to be a mistake because 'the forces arrayed against it were very strong indeed' — as if those forces were above the law, or ought to be. They would include the oil companies, the Road Haulage Association, the Society of Motor Manufacturers & Traders and, of course, their apologists — the AA and RAC. The self-interested opposition of these organisations appears to have been the reason the Conservative Government abandoned the idea of a nationally integrated transport system run for the benefit of the entire country — an ideal which even Lennox-Boyd believed 'had considerable merit'. Nevertheless, he felt that 'The Railway and the Road Haulage Executives are very separate bodies indeed and it would require a very powerful British Transport Commission to produce an integrated whole out of them' and stated that 'the BTC never had among its members anyone who is experienced in road haulage'. However, the Conservatives appear to have felt no compunction in installing non-railway people on the British Transport Commission; indeed, it could be argued that it was their specific intention to do so.

The main principle of the 1953 Act was that (to quote from internal Government memos written in 1952 and initialled by Winston Churchill) 'road haulage should be allowed to expand to the extent demanded by trade, industry and agriculture and the railways should effect such economies as they can to off-set the resulting loss of traffic and so far as they cannot do this their losses should be made up by the Transport Levy'. The latter was not a serious suggestion; the Levy was established only to compensate the BTC for the disturbance costs it would incur while it was selling-off its Road Haulage Executive.

Roads, lorries, cars and oil were to be the hope of the future, and the railways would have to sink or swim under the onslaught from those with interests in roads, contracts and cars: any money 'given' by the Treasury to railways had to be repaid with real money at interest: money spent on roads had no-one to repay it to the Treasury — road users paid a little tax, and that was the end of it.

The sale of the Road Haulage Executive was effected by a Board of six men, all appointees of the pro-road Government and all but one from the road haulage industry. It was supposed to guarantee the BTC a fair price for its assets, which all buyers knew BTC was obliged by law to sell — hardly the free market so beloved of the Government. Ministers had no clear idea of the consequences to the railway or the country of unleashing road haulage but nevertheless contemplated the coming massacre with equanimity:

'It is not possible to say how far the process will go but it might result in far-reaching modifications to the railway system. The expansion of road haulage will look after itself through modifications to the licensing system and we need not worry further about it. The achievement of economies in the railway system is a very different matter and here we may expect to be confronted with resistance from the railways. We have endeavoured to meet this possibility by widening membership of the British Transport Commission to include persons who will have an interest in seeing that these economies are effected.

'The political stake of the Government in the success of the plan is too great to allow it to be entrusted to any body of people — however able — who have no direct responsibility to the MoT or Parliament.'

The Editor and Deputy Editor of *The Economist* wrote an article criticising the proposal to sell the Road Haulage Executive whilst leaving railways under Government control and on 16 June 1952 were summoned to the Ministry of Transport for a 'rocket' from the Minister. Afterwards, the Minister wrote to the Prime Minister, Winston Churchill: 'I have explained our proposals but whether there will be any change in the tone of their scribblings [*sic*] ...'

Section 21 of the 1953 Act freed the railways from some restrictive 19th-century legislation, but Section 16 abolished the central direction of the Railway Executive, run by railwaymen, such as the great locomotive engineer, R. A. Riddles, and established quasi-autonomous Area Boards. Having broken up central control, Sub-section 16/3 demanded that the BTC establish authorities to co-ordinate the activities of all these authorities — but any new scheme or plan was liable to be 'revoked or amended by the Minister'.

The Government was aware of the likelihood of its Act's 'far-reaching' effects on the railways but was not concerned — the railways were expendable and roads would carry everything. Car and lorry manufacturers and the oil companies were the powerful forces ranged against railways, while private contractors looked forward to receiving public funds for building thousands of miles of new roads for the Ministry of Transport. Only privately owned companies were to be assisted with public funds; the railways were held down and held back.

In May 1952 Lennox-Boyd was asked in the House of Commons by Joe Sparks, Labour MP for Acton, why British Railways was so short of steel that it had to place orders for wagons with private wagon builders — which had an abundance of steel and needed to put it to use — leading to the unemployment of railwaymen wagon builders. Lennox-Boyd replied that it was normal for railway workshops to contract out wagon building and the private firms 'were subject to the same procedures for the allocation of steel supplies as the railways'. Sparks returned to the question. 'How does the Minister account for the fact that private firms have a sufficient supply of steel to take on quite abnormal and exceptional orders from BR while railway workshops are underemployed due to a shortage of steel?' Lennox-Boyd simply denied that this was a fact.

The road hauliers, well subsidised, running lorries which were frequently found to be unroadworthy and were frequently driven by men working dangerously long hours, were allowed to undercut the properly regulated railways.

The decentralisation of the Regions in the 1953 Act led to an unnecessary variety of Regional variations in equipment specifications for components in locomotives, wagons, braking, signalling and electrification equipment when the 1955 Modernisation Plan was implemented. There were six Regional Headquarters to be consulted by BTC and their compliance obtained instead of the Railway Executive handing down the standard instructions to everyone.

Two years after the passing of the 1953 Act the railway dropped below the break-even line and began to make the losses for which it was so unjustly blamed for the next 45 years.

The Modernisation Plan

Government legislation had acted as a brake on railway development for a very long time. The Railway Rates Tribunal established under the 1921 Railways Act prevented the LMS from electrifying parts of its route, even though electrification was widely regarded as desirable. Government control of railways after 1921 was strict, through the Government-appointed Railway Rates Tribunal. Each company was obliged to prove annually to the RRT that it was being run with efficiency and economy. The Government kept a control in the management in this way — although it had never contributed a penny towards railways. What was efficient and economic in the eyes of expert railway managers was not that in the eyes of an amateur. Lord Stamp, President of the LMS and a great railwayman and statistician, told his shareholders in 1935:

'There does not appear to be any likelihood of any further large scale outlay in the immediate future. We have a statutory obligation to show annually to the Railway Rates Tribunal that our affairs have been carried out with efficiency and economy and any new outlay for electrification must comply with that test.'

In 1955 the Conservative Government took credit for the intention to modernise British Railways — but put no money into the scheme. The Government had just de-nationalised road haulage, fragmented and revolutionised the organisational structure of the railway, and, while the railway officers were coming to terms with this, they were told to modernise a railway whose future size they did not know and whose traffic would be lost to competition from road hauliers beloved and pampered by Government. The Government was keen to see road haulage expand at the expense of rail, yet the Government required the whole network to be modernised. As ever, British Railways had to pay for the work out of revenue and a Government loan of £250 million which had to be repaid with interest — this on top of the millions being paid in interest to the erstwhile shareholders. And governments had the audacity to say that BR 'did not pay'! All it ever did was pay — and hold down its charges at governments' behest to make political capital for the parties.

Modernisation was to take place over a 15-year period and was estimated to cost £1,660,000,000. The interest payment on that would be £35 million per annum, and that was more than the railway could repay in the years while the work was being carried out, before the benefits of modernisation returned to raise income. As the Modernisation Plan commenced the wage bill increased, putting the railway into deficit because the Government would not allow BR to increase its freight charges — except by half what was needed — and refused to allow any rise in passenger fares. The loss to the railway of these refusals was estimated by the Ministry of Transport at £8.4 million and by the railway at £17 million a year. From 1958 the financial position of the railway went into a steep decline, with a deficit on working costs of £100 million — before the interest payments to shareholders.

Right: Although the Modernisation Plan envisaged the wholesale replacement of steam by diesel and electric traction, British Railways continued to construct steam locomotives right through until 1960. Many of the 999 BR Standard locomotives constructed between 1951 and 1960 saw less than a decade of service, as reduced traction requirements post-Beeching resulted in the much earlier-than-planned elimination of main-line steam. Nevertheless, given the potential working life of the steam locomotive, the Standard designs were barely run-in by the time that they were withdrawn (although, ironically, a number outlived some of the diesel designs that were theoretically to replace them). Two of the 251-strong '9F' class are seen at the foot of the Lickey Incline at Bromsgrove on 31 May 1963. *D. H. Cape*

Left: In the early 1950s, in an era when certain branch lines were perceived to have a future, the station at Portishead was rebuilt. This official photograph, taken on 30 December 1953, depicts the station colonnade. *Ian Allan Library*

Right: From the earliest days, the railway industry was designated as a 'common carrier', imposing upon it a duty to carry freight of whatever sort from one location to another. This resulted in most stations being provided with goods facilities and with pick-up freights running to serve them. In the early 1950s, with the rise of the road haulage industry, the railways were relieved of this duty. However, the 1955 Modernisation Plan envisaged that the railways would continue to provide a catch-all freight service, and to cater for this a network of massive new marshalling yards was proposed, with construction spanning the next decade. One of the sites selected was Millerhill (just south of Edinburgh on the Waverley route), seen here in April 1964 shortly after opening. By this date, wagon-load freight was already in decline, and Dr Beeching in his infamous report foreshadowed the demise of much more. Thus the yard at Millerhill and others such as Healey Mills and Tinsley rarely fulfilled their purpose, as later efforts to reinvigorate wagon-load freight — notably Speedlink and Enterprise — would prove ineffective. *A. A. Vickers*

Dieselisation

Modernisation required electrification where traffic density justified it and diesels for everything else — and as quickly as possible. Railway engineering matters had evolved steadily over the previous 120 years developing robust, reliable machines by following well-understood techniques. Now everything had to be done in a hurry, with the overheated breath of Conservative politicians breathing down the necks of the conservative engineers. The old central control was abolished in October 1953 in favour of a new fragmented and Regionally-autonomous organisation which would take time to set up. Committees abounded. Meanwhile a temporary organisation was established, in place by 1954, and at the start of 1955 this was abandoned for the final version. The Regions at once started to do things their way, with much arguing with Headquarters, and it was in this atmosphere that the planning of the modernisation of the railway took place.

Meanwhile the two excellent English Electric-engined Co-Co diesel-electrics built by the LMS and the pair of similarly powered 1Co-Co1s built in 1950/1 by the Southern Region had been doing good work. From 1953 all four were allocated to Camden on the LMR for top-link workings, and in 1955 a fifth SR machine joined them. Why not build on that experience in consultation with English Electric? It was a good system from a good manufacturer and indeed, the diesel-electric system (a diesel engine driving an electric generator) was — after a great deal of money had been spent — accepted as standard.

Responsibility for 'dieselising' the railways of Britain fell to a team led by Deputy Chief Engineer Roland Bond, steeped in the mechanics of the steam engine. There was not, in Britain, a great deal of experience in operating main-line diesel locomotives, but one thing was known: the best policy would be to standardise on a basic design of locomotive — with whose controls all drivers could become familiar — with as few variations (in terms of size and power rating) as possible.

Fell diesel-mechanical No 10100, pictured at Derby on 21 April 1955. This locomotive had been built as a private venture in 1950 and was acquired by BR in 1955 after it had been out of service for repair for about a year, (during which time this photograph was taken) and survived until 1960. In tests on the LMR it proved successful, but by this stage BR had decided to concentrate on either electric or hydraulic transmission. *P. H. Groom*

The four-wheel diesel railbus was perceived to be the saviour of the lightly trafficked branch line, and a number were supplied by a variety of manufacturers. This is a Park Royal-built example, seen at Ayr with a service from Kilmarnock on 22 June 1962. Despite the provision of additional stations designed to encourage local traffic, the railbuses proved unsuccessful. *S. Creer*

The Pilot Scheme

In 1955 BR ordered 174 new locomotives from six manufacturers — Birmingham Railway Carriage & Wagon Co, British Thomson-Houston Co, Brush Traction Ltd, English Electric, Metropolitan-Vickers (AEI) Ltd, North British Locomotive Co — plus BR's own Works at Derby. Power units came from seven manufacturers — Crossley, English Electric, MAN, Maybach, Mirrlees, Paxman, Sulzer. Besides these were a score of manufacturers of ancillary equipment — Self-Changing Gears (transmission), Serck (radiators), Spanner (boilers, necessary to ensure the new traction was compatible with steam-heated carriage stock), Stone (boilers) and many more. The permutations of equipment were endless. There was no home market for diesel engines if BR did not buy them, and without the basic support to their finances of building for the home market they could not make the extra numbers for export. The Government was keen that BR should buy a varied selection of diesels to give British industry an opportunity to 'showcase' its products — but gave BR no grant aid with which to buy diesels. British Railways needed standardisation of locomotive components for cheapness of purchase and standardisation of controls to make them universally driveable by any crew anywhere in the country, but this was not going to happen while BR it was having to assist the business of other companies. When some of these hastily devised and untried machines proved to be useless, BR carried the cost — and the curses of the Government for losing taxpayers' money! Governments were never behind hand in their criticism of the 'incompetence' of a 'state-owned monopoly'. The Conservative Government said it wanted BR to be a free-standing, profit-making organisation, yet the railway was not allowed to follow its own best interest but — as had been the case since 1921 — to be used as a conduit for supporting private companies.

The 174 diesel locomotives ordered in 1955 fell into one of three horsepower groups — 800-1,000 (Type A; later Type 1), 1,100-1,250 (Type B; later Type 2) and 2,000+ (Type C; later Type 4) — and were intended to take part in the Pilot Scheme, the idea being to allocate them to certain main-line depots and monitor their performance carefully before any more were purchased. However, the 18 months it would take to build them and the 18-month trial period meant three years of non-equipment with further examples of these life-saving (for BR) machines; with the railway's financial position deteriorating rapidly, the BTC decided that the that trial periods were a luxury it could not afford. In 1956 BR fell below break-even point, and in 1957 there commenced a mass ordering of diesel locomotives commenced. In theory these could work 22 hours a day, requiring minimal maintenance, and thus fewer of them would be needed to maintain the same level of service. The result was the introduction of large numbers of untried diesel locomotives, many of which were soon found to be wholly unsuitable for the tasks expected of them and which in many cases barely outlasted the steam locomotives they were purchased to replace. As if this were not wasteful enough, between 1956 to 1960 new steam engines were still being built in the railway workshops, alongside the new diesels. The new steam engines were redundant even as they were being built, and a large number of them never worked long enough to wear out their boilers.

The first diesels to be introduced under the Pilot Scheme — the 1,000hp English Electric Type 1s (later Class 20) — were, relatively speaking a runaway success, but in 1962, after 128 had been turned out, production was curtailed in favour of a new design — the Clayton Type 1 (Class 17) — featuring twin Paxman engines developing a combined total of 900hp. In all, 117 engines of this class — Nos D8500-8616 — were purchased by BR, the last not appearing until 1965. As early as 1963 it was realised that they were a disaster, but BR continued to buy them; doubtless there was a contract, but a trial period would have been a good idea before signing up for 117 locomotives. The power units suffered from cylinder-head failures and broken crankshafts, and by the mid-1960s no less than 40% of the 'Clayton' fleet was under repair as a matter of course. Finally, in 1966, common sense prevailed, and BR returned to English Electric for a further 100 of its own Type 1 design.

The 'D82xx' Type 1s (later Class 15), introduced from November 1957, featured 800hp Paxman engines and BTH generators and were reasonably decent (if somewhat puny) machines. However, to please GEC, the broadly similar 'D84xx' (Class 16), which followed from 1958, had GEC generators and were poor tools; after fairly non-productive lives, including many

months in works having engines changed after overheating and seizures and the failure of their electro-magnetic control systems, they were withdrawn in 1968.

The 'D61xx' Type 2s (later Class 21) were a maintenance disaster built by the once great North British Locomotive Co, whose steam engines ran all over the world, in war and peace, with enormous success. Fitted with 1,100hp MAN engines, after trials all 10 of the original batch were sent to Hornsey during January 1959. A year later they were all in trouble with frequent failures of all sorts but in particular of the coupling between the engine and the generator. In September 1960, with their engines derated to deliver 1,000hp, all were reallocated to Scotland, where they were found light duties and were conveniently close to their manufacturer's workshops. Twenty (subsequently Class 29) were later re-engined with 1,360hp Paxman units, and, although these represented an improvement, they did not last significantly longer than the unmodified locomotives, all being scrapped by the early 1970s. Broadly similar to the 'D61xx' were the 'D63xx' (Class 22),

The Class 17 Clayton-built Bo-Bos ultimately numbered 88, a further 29 being constructed by Beyer-Peacock. Production of the class ran from September 1962 to April 1965; unfortunately, the type of traffic for which the class was designed — secondary passenger and pick-up freight — were exactly the types of traffic identified by Dr Beeching for

the axe, so that many of the type saw only four or five years of service. The last of the class were withdrawn in 1971, although one survives in preservation. Here, on 25 October 1962, No D8505 is seen at Tarbolton with the 1.30pm Falkland Junction (Ayr)–Saltney (Chester) fitted freight. *Derek Cross*

Well-established and successful as a manufacturer of steam locomotives, the Glasgow-based North British Locomotive Co came unstuck when it turned to the production of diesel locomotives. Amongst its failures were batches of Bo-Bo/B-B classes supplied to Scottish and Western Regions respectively. The ScR examples were provided with electric transmission, those on the WR with hydraulic. No D6100 was the first of the ScR examples and was built in December 1958. A total of 58 locomotives of each type were supplied between 1958 and 1962; the majority saw less than a decade of service. *Ian Allan Library*

all allocated to the Western Region. However, the WR alone opted for hydraulic transmission — non-standard for British Railways and creating yet more variety in its diesel fleet.

The 'D59xx' Type 2 'Baby Deltics' (later Class 23) were from the English Electric stable — a mark of good breeding. They were fitted with an 1,100hp version of EE subsidiary Napier's 'opposed piston' Deltic two-stroke engine, which should also have been a guarantee of success, and were sent new to King's Cross to work on inner- and outer-suburban trains. The strain proved too much for their engines, which were in serious trouble within weeks, obliging BR to relegate the locomotives to shunting and local goods work. They were duly sent back to English Electric at Vulcan Foundry, were they were rebuilt in the course of a year. They went on to do good work, but withdrawal began in September 1968, and the last example was retired in March 1971.

Remarkable for having one six-wheel and one four-wheel bogie, the Metropolitan-Vickers Type 2 (later Class 28) was powered by a Crossley two-stroke engine which gave nothing but trouble. The first, No D5700, began trials in July 1958, but the problems seem to have concealed themselves until the engine had finished its trials and was allocated to Derby to work main-line passenger trains; all 10 were at work by October 1959. Arguably, they should never have been used in Britain at all, since a very similar design from the same manufacturer was already making itself useless in Ireland. By February 1961 the entire class was stored awaiting repairs by manufacturer, but all were back at work by October that year. They were sent to

Barrow (as opposed to Coventry), but their diagramming was such that a pair of them had a regular job on the 'Condor' express freight, double-headed from Hendon to Glasgow — 404 miles. However, despite the rebuild they remained beset with serious problems, and withdrawal began in 1967, the last of the class departing unlamented in December 1968.

The Brush Type 2 (later Class 30) was the second Pilot Scheme design to be introduced, late in 1957, and by the end of 1962 no fewer than 263 had been built. Only later was it discovered that the crank case of the type's 1,250hp Mirrlees engine — a power unit designed for marine applications — was distinctly fragile. Having purchased so many examples, BR had little choice but to persevere, but when modifications proved unsuccessful it finally bit the bullet and had the entire class re-engined by English Electric (Class 31).

As delivered, the least troublesome Type 2 diesels were probably the numerous Sulzer-powered locomotives from BRCW (Classes 26 and 27) and BR's own Derby Works (Classes 24 and 25). Altogether there were over 1,200 Type 2 diesels; indeed, it seemed as if the ghost of the Midland Railway hung over BR, such was the apparent popularity of puny locomotives. This turned out to be a double blunder: already underpowered, as branch and secondary lines closed the Type 2s started to receive a hammering on more arduous duties, increasing their failure rate. Of the Pilot Scheme diesels, only the heavy English Electric and BR/Sulzer Type 4s (later Classes 40 and 44 respectively) can be said to have been directly descended from the LMS and SR designs of the early postwar period.

Above: The 10 'Baby Deltics' were all delivered in 1959 and were designed primarily for use on suburban services out of King's Cross. No D5904 is pictured in 1967 less than two years before withdrawal. *Alec Swain*

Below: One of the Metropolitan-Vickers Co-Bos, No D5700, at Derby on 28 July 1958 when brand-new. A total of 20 of the class were produced between July 1958 and October 1959; all had been withdrawn by December 1969, many having suffered longish periods in store during their careers. *A. N. Yeates*

'Warships', 'Westerns' and 'Hymeks'

In 1956 the Western Region was under the chairmanship of Reginald Hanks, a pre-1914 Swindon Works apprentice who had served in the Great War and then gone to work making cars at Morris Cowley. He was delighted be in charge of his beloved 'Great Western' and regularly fired the 'Castle' hauling his train from Oxford to Paddington on his way to work. He told me that he had 'no idea' what sort of diesel he wanted for the WR, but, noting how well the Deutsche Bundesbahn 'V200' diesel-hydraulic performed and its very favourable compactness, power-to-weight ratio and speed, that would be the engine for him. Each locomotive would cost less than a 2,000hp diesel-electric, and because of its excellent power-to-weight ratio, would be able to haul two more coaches at the same speed. The savings were impressive, and the British Transport Commission agreed to let the Western go its own way. The WR version of the German 'V200' weighed 79 tons against the monstrous 133 tons of the English Electric Type 4 (Class 40).

No D800 *Sir Brian Robertson* (named after the Chairman of the British Transport Commission) appeared on 3 June 1958 — the first of 71 B-B 'Warships' built between then and 1962. Its twin high-revving Maybach 650 power units each drove through a Mekydro hydraulic transmission and together developed 2,100hp; later examples developed 2,200hp, while others had MAN engines and Voith transmission. However, the capital value of the German spares that WR had to carry was three times greater than that of, say, EE spares. The mechanisms were very complicated, difficult and expensive to maintain — and they broke. The transmission was relatively fragile compared to the ruggedness of the diesel-electric, and it was the transmission problem that finally put paid to the 'Warships'. In Germany the 'V200s' were very successful but were not normally driven beyond 75% — 80% at most — of their full power. On the Western Region the 'Warships' were expected to produce 100% power for lengthy periods — there is no point in paying for 2,000hp and then using only 1,500hp. This was the British and American view generally, but perhaps, even in diesel engines, a little mercy can often save a lot of trouble. Whilst it cannot be denied that the 'Warships' did some very fine work, even beating the best speed of the 'Castle' steam engines on the 'Bristolian', they were being thrashed hard to beat the best that WR steam could offer, and that was not good for their transmissions; the 'Warships' were so failure-prone that the WR actually considered replacing them before they were life-expired in accounting terms.

In 1959/60 a more powerful type of diesel-hydraulic was designed for the Western Region. The first of this class, No D1000 *Western Enterprise*, entered traffic on 20 December 1961, just as the last of the 'Warships' were introduced. Swindon and Crewe works were given orders to build 74. The 'Westerns' were powered by two Maybach 655 engines to give 2,700hp. They drove through a Voith hydraulic transmission and weighed 108 tons. The first

Between 1958 and 1962 the WR took delivery of 71 'Warship' diesel-hydraulics to a design derived from that of the Deutsche Bundesbahn 'V200' class. Maybach-engined No D816 *Eclipse* stands at Paddington on 10 September 1960. *D. C. Ovenden*

Above: Although the 'Westerns' eventually proved themselves to be the most successful and long-lived of the various diesel-hydraulic classes, surviving until the mid-1970s, their early years were problematic. On 18 October 1962 No D1035 *Western Yeoman* approaches High Wycombe with the 11.40 service from Birkenhead to Paddington. *H. K. Harman*

'Westerns' were less than six months old when they had to go back to Swindon to have their suspension modified to cut out a dangerous 'roll' which took place when passing over pointwork at speed — a problem shared with the Brush Type 4 diesel-electrics. At about six months old, having run about 100,000 miles, the 'Westerns' suffered transmission failure because of a roller bearing that broke up. Each bearing took two days to replace, but the replacement bearing was the same inadequate design and broke up after a further 100,000 miles. The 'Westerns', which had been running on the South Wales route, were withdrawn (temporarily) in batches; they were replaced with 'Hymeks' used hitherto on Paddington–Worcester expresses, onto which now came back the much-maligned steam engines — including No 4089 *Donnington Castle*, which looked as if it had come straight from the scrapyard.

Introduced in 1961 and built by Beyer-Peacock, the 'Hymek' was a diesel-hydraulic design powered by a single Maybach engine of 1,700hp — as much as a single hydraulic transmission could then handle. These locomotives nevertheless found the Worcester-line work a great strain (as they had the Gloucester–Swindon run) and broke down frequently. Deemed capable of working 22 hours a day, a single locomotive would be diagrammed to work several trains, but this meant that, when it broke down, several trains were lacking motive power. At Swindon there was one which had failed and was out of traffic for weeks because the fault could not be found. The fault was eventually traced, and the Locomotive Foreman wrote furiously in the 'Faults/Remedies' book: 'Cause of Failure — Cup of Tea'. Cups filled with sugary tea were left standing on the cab console, and, as the engine sped along, the tea slopped over, carrying sugary solution through to the underside, where electrical contacts became coated with insulating (and invisible) sugar. To men brought up on the ruggedness of the steam locomotive, the idea that a locomotive could be brought to a stand by a slop of tea was outrageous. But, of course, these men were dinosaurs and would, in a few years, follow their steam engines into extinction, leaving the railway to its bright new future.

Built by Beyer-Peacock, the 101 'Hymek' diesel-hydraulics were delivered between 1961 and 1964. At one stage a further 200 were envisaged, but in the early 1960s the BTC decided — somewhat belatedly — to standardise on diesel-electrics. No D7093 is seen at Swindon shed on 20 September 1964.
Brian Stephenson

The life of the BR Type 1 0-6-0 (Class 14) diesel-hydraulics — nicknamed 'Teddy Bears' — was certainly a picnic as far as their operation on the Nationalised railway was concerned. Introduced to the Western Region in July 1964, the class numbered 56 locomotives constructed at Swindon Works by October 1965; however, the traffic for which they were designed — primarily pick-up freights — had been largely abandoned as a result of the Beeching Report, and the entire class had been withdrawn by April 1969. In the worst case, No D9531, its BR career lasted only from February 1965 until December 1967. The vast majority, however, were sold to industry both in Britain and overseas, and almost half now survive in preservation. No D9528 is pictured at Bristol on 20 March 1965.
R. A. Panting

Troublesome Traction

Despite the fact that Britain has been at the forefront of railway technology for two centuries, the history of the development of locomotives and rolling stock has not been universally successful. Throughout the book, a number of examples of failed developments have been linked to individual engineers, but this section provides an overview of many others. No doubt there are many more — indeed, each individual enthusiast will doubtless have his or her own hit-list of the industry's spectacular engineering own-goals — but this selection may be regarded as including some of the most notable.

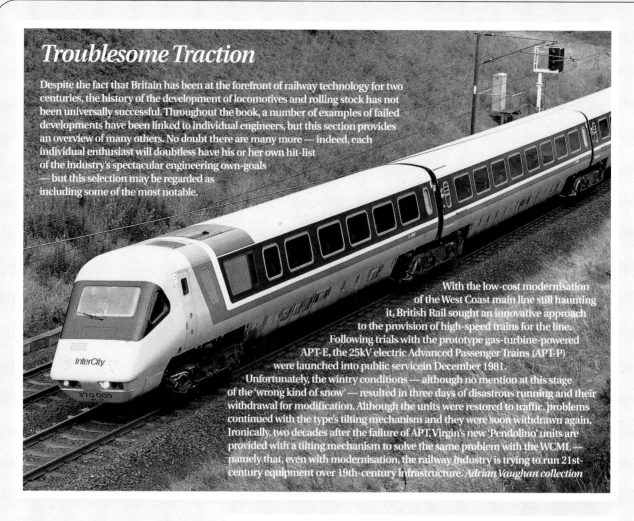

With the low-cost modernisation of the West Coast main line still haunting it, British Rail sought an innovative approach to the provision of high-speed trains for the line. Following trials with the prototype gas-turbine-powered APT-E, the 25kV electric Advanced Passenger Trains (APT-P) were launched into public service in December 1981. Unfortunately, the wintry conditions — although no mention at this stage of the 'wrong kind of snow' — resulted in three days of disastrous running and their withdrawal for modification. Although the units were restored to traffic, problems continued with the type's tilting mechanism and they were soon withdrawn again. Ironically, two decades after the failure of APT, Virgin's new 'Pendolino' units are provided with a tilting mechanism to solve the same problem with the WCML — namely that, even with modernisation, the railway industry is trying to run 21st-century equipment over 19th-century infrastructure. *Adrian Vaughan collection*

The Class 303 EMUs were introduced to service on Glasgow's northern suburban routes in November 1960; however, within a month of their introduction, the type had had to be withdrawn after a number of transformers on the units blew up. Steam suburban services were reintroduced, and it was not until towards the end of the following year that the class was reinstated following modification. This was not the only weakness evinced by the type; the curved windscreens were also problematic in that it was difficult to keep them watertight, and these were replaced with smaller flat glass panels. This view shows one of the units under test in August 1960 at Craigendoran. *British Railways*

Right: Whilst the Brush Type 2 (today's Class 31) diesel-electric has ultimately proved to be one of the most successful and long-lasting of the Modernisation Plan designs, its early life was not without problems, largely as a result of the Mirrlees diesel engine originally fitted. This proved prone to fatigue, particularly at higher power ratings, with the result that, from 1964 onwards, the class was re-engined with English Electric units; the original engines were returned to Mirrlees and reconditioned for use in trawlers. Before the problems emerged, Mirrlees boasted proudly of its involvement in the Modernisation Plan, as evinced in this contemporary advertisement featuring No D5500, the first of the class. *Ian Allan Library*

Below right: Numerically the largest class of main line diesel-electric to be constructed for use on British Rail, the Brush/Sulzer Type 4 (later Class 47) eventually numbered 512 locomotives. Despite the numbers built, however, the class was not without its tribulations. The Sulzer engines required modification, which resulted in the class being derated from a notional 2,750hp to 2,580. Five of the class (allocated Class 48 under the TOPS scheme), Nos D1702-6 (including No D1704, seen here at Crewe on 6 June 1967), were fitted experimentally with an alternative Sulzer engine which ultimately produced less power, and the five were provided with standard Sulzer engines in 1969. Another problem was that BR decided to perpetuate the use of steam heating of carriage stock, with the result that many locomotives were not constructed with electric train-heating equipment; this was later fitted to many of the class but had the effect of reducing power. Although many of the class have seen more than 30 years of service — indeed, some have now entered their fifth decade — problems with bogies and motors ensure that constant vigilance in maintenance is necessary.
David Wharton

Left: Promoted in the mid-1980s as a means of encouraging the regeneration of London's deprived docklands, the Docklands Light Railway would ultimately be both highly successful and a victim of its own success. However, the early years proved problematic, as the computer-control system proved incapable of operating the automatic trains correctly, there being countless instances of trains stopping between stations and missing legitimate stops, with the result that the 'train captains' were invariably to be found taking manual control. The first day of public service, 31 August 1987, perhaps provided a portent of the future when No 10, pictured here at Stratford, had to be re-routed at West India Quay as a result of the failure of unit No 8 at Westferry, the latter being rescued by No 3. Fortunately, the teething troubles proved soluble, although the original batch of vehicles was replaced quickly and sold to Essen, Germany. Today the DLR continues to expand and plays an important role in the provision of public transport in the area.
Michael McGowan

The SMJ

There was always a dearth of east–west railway routes in Britain, and from time to time in the 19th century some optimistic contractor would try to construct one. Several little companies developed parts of such a route across the middle of England. They had hopes of iron-ore traffic locally and traffic from the great companies they crossed, but they led a precarious and sometimes bankrupt existence. In 1908 they agreed to amalgamate for added financial strength. Thus was formed one larger, struggling company — the Stratford-upon-Avon & Midland Junction Railway (SMJ). Its route covered about 85 miles as an energetic crow might fly and nearer 100 miles as the meandering train would puff — from Ravenstone Wood Junction, three miles north of Olney on the Midland Railway's Bedford–Northampton line, to Broom Junction, on the Midland's Ashchurch–Redditch line. North- and southbound Great Central

trains could run direct onto the westbound SMJ at Woodford Halse. There were indirect junctions with the LNWR at Blisworth on the Euston–Birmingham line and with the GWR's Banbury–Birmingham line at Fenny Compton. The SMJ had its own station at Stratford-upon-Avon and crossed the GWR Birmingham–Cheltenham line there on a bridge. A curve from the SMJ led northbound into the GWR line. Beyond Stratford-upon-Avon the SMJ met the Midland at Broom with forked junction, turning north and south. Both the Midland and the Great Central had running powers over the whole of the SMJ: from 1908 the Midland used it for its London–Avonmouth/Bristol freight, avoiding an even longer journey through Birmingham (and reducing congestion there); the GCR put East Midlands coal on it. However, in 1913 both companies withdrew their patronage, and the SMJ lost £2,700 in mileage payments. Possibly the two

Class 4F 0-6-0 No 44186 moves its light load after the stop to collect the train staff at Ettington in early 1955.
Donald Kelk

The heyday of freight on the SMJ: ex-GWR 'Hall' 4-6-0 No 5990 *Darford Hall* and an LMR Class 8F wait while Class 9F 2-10-0 No 92246 brings a South Wales iron-ore train off the single line from Fenny Compton. *John R. P. Hunt*

larger companies hoped to bankrupt the SMJ by starving it of traffic, whereupon they could buy it jointly at a knockdown price, but World War 1 intervened, bringing the SMJ under Government control and providing some extra traffic — although, of course, it was paid its 1913 revenue by the Government, so the extra traffic was carried for nothing.

The SMJ became part of the LMS in March 1923, whereupon the latter's operating authorities resurrected the Midland Railway practice of running serious freight between St Pancras and Avonmouth/Bristol over the SMJ. In 1926 Sir Josiah Stamp, Britain's greatest statistician and Chairman of the LMS, authorised expenditure on upgrading to main-line standards the SMJ's track from Ravenstone Wood Junction to Broom Junction. The work was carried out in 1927/8, after which LMS freight between London and Bristol passed over this remote but upgraded byway. During World War 2 this east–west route and its connections to north–south main lines was again of great value, and in 1942 this value was increased by the construction of a curve from the SMJ to the southbound Redditch–Ashchurch line.

The one great selling point of the line for passenger traffic was Stratford-upon-Avon, where in 1931 the LMS had opened the Welcombe Hotel; the following year the railway began running a petrol-engined road/rail bus from the Euston main line at Blisworth to the SMJ goods yard at Stratford, where the rail wheels were retracted and the rubber-tyred road wheels lowered so it could drive through the streets of the Bard's birthplace to the

Welcombe Hotel. This most enterprising service lasted only weeks.

During the two World Wars the SMJ had demonstrated its strategic value in carrying iron ore between the Midlands and South Wales, the ironstone fields of Wellingborough being well placed to make use of the line. The alternative was a fantastic detour south to Bletchley, west to Yarnton, northwest to Honeybourne and thence southwest to South Wales. There had also been ironstone quarries on the SMJ itself, at Edge Hill, in which money was invested during World War 1, but the venture did not survive the war.

When railways were nationalised in 1948 the SMJ became a national resource rather than one belonging to a particular company, and the future looked bright for the men who worked it. However, it was never successful as a passenger railway, and its passenger services were all withdrawn by April 1952.

In 1948 several privately owned, South Wales-based steel-making companies were amalgamated to become the Steel Company of Wales (SCW), which duly began to plan a major expansion of existing plants at Margam, a few miles south along the coast from Port Talbot. Some of the iron ore for the new works would be imported, but at that time the sea-lock of Port Talbot could not take ships of more than 5,000 tons deadweight. The cost of shipping high-grade ore in such small quantities from the other side of the world was high, and ore from the Irthlingborough and Banbury quarries would be required in very large quantities once the enlarged steel works were opened.

The Banbury ironstone came from the Oxfordshire Ironstone Co (OIC) quarry at Wroxton, developed in 1917 in response to an increased demand for home-produced steel. OIC money was poured in, and a railway, six miles long, five of which were single-track, was constructed — to a very high standard, complete with signalboxes — down to the GWR main line at Ironstone Sidings, $1\frac{1}{2}$ miles north of Banbury station. By the time construction was complete the war had been over for three months. The quarry produced 112 tons of stone in 1919 and 600,000 tons in 1929, following which the steel industry began its descent into recession until rearmament in 1935; by 1939 production was in full swing and rose annually thereafter.

Around 1951 vast reserves of high-quality iron ore were discovered in Labrador, South America and Australia. In the vast wastes where this ore lay, it could be ripped out by gigantic machines, making the methods of extraction at Wroxton — a few Ruston Bucyrus steam shovels — look puny. The virtually unlimited production potential of Western Australian quarries demanded 100,000-ton ships, and if these could enter Port Talbot Harbour this vast production could be delivered to Margam at less per ton than the cost of extraction at Wroxton. Moreover, Wroxton ore had an iron content of only 30% — half that of the Australian ore. The SCW therefore began to investigate a major enlargement of Port Talbot Harbour.

In 1953 the Oxfordshire Ironstone Co's railway line was doubled throughout, and in 1956 the quarry produced $1\frac{3}{4}$ million tons. The Western Region was hauling all this output to Margam via Leamington, Hatton Bank and, turning south at Hatton, all the way downhill to Stratford-upon-Avon and South Wales. Compared to the SMJ route from Fenny Compton to Stratford, this represented a 27-mile detour. In 1950 control of this section of the SMJ had passed from LMR to WR, which in 1954 decided to spend a large amount of its money on upgrading the SMJ between Fenny Compton and Stratford.

After four years of planning, the rebuilding commenced. The layout at Fenny Compton was transformed with a junction facing Banbury and up and down reception lines to accommodate ironstone trains awaiting their path; this required a new, 77-lever signalbox. At Stratford-upon-Avon, a double-track curve east to the

On 20 June 1964 Class 9F 2-10-0 No 92213 heads towards Woodford, near Fenny Compton, with a Class 7 freight. *Gerald T. Robinson*

In March 1964 Class 8F No 48475 climbs away from Ettington towards Stratford with a freight train. *T. E. Williams*

southbound Birmingham–Cheltenham line was constructed. This required the demolition of the old SMJ engine shed and the design and construction of a 50-lever signalbox — Evesham Road — to control all the tracks at what had now become a triangular junction; the reader will recall that there was already an east–north curve here. Eight miles further south, at Honeybourne West Junction, locomotive and crew changes were to take place. Three long loops were installed, along with facilities for locomotive watering — a major operation in its own right — mess rooms and a new signalbox with 50 levers.

The work was brought into use in stages, culminating with the commissioning of the Fenny Compton installation on 12 June 1960, simultaneous with completion of the expansion of the Margam steel works. By this time, however, ironstone production at Wroxton had entered into decline: in 1959 the quarries were on a three-day week, and production never again reached 1956 levels. On a good day in the early 1960s just six or seven 14-wagon trains were brought down from the quarry to the WR main line at Ironstone Sidings; from there they went on to Margam, via the SMJ, hauled by BR '9F' 2-10-0s and 'WD' 2-8-0s. But the Western Region raised its charges for carrying the ironstone, perhaps to recoup the capital

it had invested in the refurbished SMJ. In 1961 the Oxfordshire Ironstone Co decided to reduce its costs by investing in two 0-4-0 Sentinel diesel-hydraulics as a trial with a view to replacing all 22 steam engines.

By June 1964, although there was still 40,000 tons of ore a week to be moved from Wroxton, there were no scheduled trains on the SMJ, the ironstone trains having reverted to the circuitous route via Hatton, where they delayed passenger trains. On 1 March 1965 the entire SMJ line was closed. The WR had pumped tens of thousands of pounds into a route which it used for barely four years.

In spite of the opening, in 1964, of a new steel-making plant at Newport — Llanwern — in addition to Margam, the demand was for imported ores. Wroxton's workforce was reduced from 190 in 1956 to 135 in 1963. The full complement of new OIC diesels began arriving in September 1964 and the 13th — and last — arrived in July 1965; by 1966 one diesel was enough for the 2,000 tons a week then being produced, and in September 1967 the quarry was closed.

The Oxford–Cambridge Line

After the war, re-housing strategies developed during the war began to be realised. Oxford, Buckingham, Bletchley, Bedford and Cambridge were all to take large increases in population, and new towns were envisaged. Great surveys were conducted, and vast plans were made for London and for the South East of England which laid the strategy for new towns and a new city. Bletchley Town was adding new houses at the rate of 200 a year in the early 1950s. In 1958 a proposal for an 80-acre industrial estate and an expansion of 23,700 houses was approved to the east of the town, close to the Bedford–Bletchley railway line. Development continued as Londoners were re-housed as a matter of Government policy.

On 31 March 1956 ex-LMS Ivatt Class 4 No 43089 passes Oxford North Junction with the 2.38pm from Oxford to Cambridge. The section of line from Oxford to Bicester Town was to be reopened for passenger services in May 1987, but the section beyond Bicester towards Bletchley and Milton Keynes remains freight-only, despite efforts to see the route reopened for passenger services. *G. D. Parker*

In 1955 the BTC, purely for operational reasons rather than in consultation with town planners, decided to increase its use of the Oxford–Cambridge line (Oxbridge line). The route provided a connection to/from all East Anglia and its ports and crossed four north–south main lines, with connections to three of them, whilst leading directly to Worcester, South Wales and the West of England via Oxford. Traffic for the Midlands and West of England, which otherwise had to go into London, could be passed quickly across country by this route, and on 29 September 1956 work began on the Bletchley flyover, to carry the east–west line above the West Coast main line, and on the replacement of the old flyover across the ER main line at Sandy. In addition, construction of a mechanised marshalling yard commenced at Swanbourne, on the Oxford line, five miles west of Bletchley.

In May 1959 BR decided that freight traffic was in terminal decline and that the Oxbridge line was no longer necessary and should be closed. This at a time when BR was ordering dozens of small diesel locomotives to haul

On the eastern section of the line, a Cambridge–Bletchley train headed by Class D16/3 No 62535 prepares to cross the East Coast main line at Sandy. The section east from Bedford to Cambridge is now abandoned, although the trackbed, as here at Sandy, is still easily identifiable. Many of the bridges have, however, disappeared. *G. W. Goslin*

such freight and when Bedford and Bletchley were still increasing their population; Buckingham, on a branch line off the Oxford–Cambridge 'bridge', was also growing, and a new garden city — for 250,000 people — had been formally proposed for an area of green fields northwest of Bletchley, centred on a village called Milton Keynes near which the Oxford–Cambridge line passed.

In spite of BR's declared intention to close the line — and of the fact that there was no question of its becoming a major east–west route — construction of the Bletchley and Sandy flyovers continued; the Bletchley flyover was duly completed in January 1962, having cost £1.5 million, yet no daily, regularly scheduled train ever used it.

Opposition to the closure of the Oxford–Cambridge route hardened with the formation of the Oxon & Bucks Railway Action Committee (OBRAC), which managed to persuade BR to introduce diesel railcars in November 1959. In 1963 Dr Beeching was unsure of what to do with the line. He seemed to recognise its strategic importance, because, rather than marking it for closure, he condemned its intermediate stations to closure and retained it for through freight, parcels and even passenger trains. But BR did not introduce faster schedules between Oxford, Bletchley and beyond — there was no positive effort made to make the line serve the community, ease the traffic jams and make life better for the constantly increasing populations of the towns along the way. Instead the train service was reduced to one of those 'linger & die' situations — a sulky service well known elsewhere.

Having throttled the line for a number of years, BR put the strangled remains up for outright closure in 1964, using financial and passenger statistics that everyone but BR felt were incorrect; in spite of the dismal train service, the line had not been doing as badly as BR claimed. The Labour Minister of Transport, Tom Fraser ('Vote for Wilson and get rid of Beeching'), nevertheless approved the closure, and the passenger service between Bletchley and Oxford and all services on the Bedford–Cambridge section of line ceased after the last train on 30 December 1967. Buses were run, subsidised by BR and the county councils.

When I was a signalman at Oxford in 1969 I noticed that there was a Paddington–Oxford DMU which stood for two hours between 8 and 10am, driver in cab, engines running, in a siding at Oxford before returning to London. I suggested that, rather than enter sidings, it could continue from Oxford station to Bicester and attract some passengers who would otherwise join the vast jam which at that time of the day built up all around the Oxford ring road and in the Banbury Road, but the idea was said to be 'too expensive' to realise. How cheap does operating have to become before it is cheap enough?

When Oxford-area signalling was automated in 1973,

Right: One of the network of lines that disappeared in Buckinghamshire in the 1960s, before the massive growth of the population in the region and the rise of longer-distance commuting, was the branch that served Buckingham itself. On 21 August 1964 Derby Lightweight diesel railcar No M79901 stands in the station with the 4.35pm departure. *Alec Swain*

Below: The only section of the line to retain passenger services is that between Bletchley and Bedford, and even this section of line has had its services threatened on more than one occasion. On 17 November 1986 a three-car DMU arrives at Lidlington with the 13.33 service from Bletchley to Bedford. *J. Critchley*

Blunder, white elephant or infrastructure constructed before its time? Undoubtedly the Bletchley flyover has rarely fulfilled the expectations of those who proposed its construction in the 1950s as part of a radical scheme for improved transport links, primarily for freight, around London. Currently mothballed, the line may yet again feature as rail-borne freight traffic — particularly containers to/from Southampton and Felixstowe — continues to grow.
On 24 August 1980 a special from West Ruislip to Grange-over-Sands, headed by Class 47 No 47173, was diverted over the concrete structure.
T. Parkins

the line to Bicester was singled but remained double from Bicester to Claydon LNE Junction. Minimal signalling was installed, insufficient for running passenger trains — the crews of goods trains had to operate the level crossing barriers along the way. In May 1987, after expensive re-signalling, BR introduced three trains a day over the 10 miles to terminate at Bicester. In May 1989 the old halt at Islip was rebuilt and a seven-train service instituted. The line east of Claydon LNE Junction became derelict. In July 1993 thieves stole the signalling cables — a determined effort, because the cables were buried and had to be dragged out of the ground by a JCB. These same criminals or some other idiots stole some of the track in the vicinity of Swanbourne, so the rest of the line, from Bletchley to Claydon LNE Junction, was lifted by Railtrack. Just beyond the buffer-stops east of Claydon LNE Junction there is an automatic open level crossing and this was still, in 2002, maintained in working order, although there are no rails — making testing of the crossing lights' circuitry difficult.

British Rail announced its intention to close the Bedford–Bletchley line in October 1971 — this in spite of the fact that it served two major towns connected by a winding and heavily congested road system and that its Woburn Sands and Aspley Guise stations were within vintage-bus-drive distance of Woburn Abbey, one of Britain's major tourist attractions. The railway's lack of enthusiasm for running railways was remarkable. Milton Keynes Development Council and others managed to keep the Bedford–Bletchley line open, so that in 2003 the line remains open for passenger trains running at a

maximum of 50mph, pending re-signalling. On the Bedford bypass cars may run at 70mph, so again it appears that railways are not taken seriously — or else they are taken so seriously (as a competitor to roads) that they have to be kept in shackles.

The SRA rejected six years of planning and feasibility studies when it rejected the proposal to re-open the Oxford–Cambridge railway as a 90mph double track because, it said, re-opening would be too expensive and would not bring about the hoped-for improvements; moreover, the track would have to be re-laid for many miles between Cambridge and Bedford via Sandy following the old alignment. The purpose of the link was to connect East Anglia with the west and relieve traffic on A14 and M25. The cost was said to be £240 million, but the SRA apparently did not think this a realistic figure. The SRA also stated that 'Railtrack has insufficient finance to work on new projects because it is trying to deliver a reliable infrastructure and competent maintenance regimes following Hatfield'.

On 14 August 2001 the Strategic Rail Authority announced the very parochial decision to abandon the Oxford–Cambridge as a through route. One day, perhaps when the newly opened Bedford bypass is solid with traffic and no more lanes can be added to the M25, the Strategic Rail Authority will appreciate the strategic importance of the Oxford–Cambridge line and re-invent it as a major rail route. And then it will be opened by King Charles III — or even King William V — as if it had been the Authority's intention all along that the line should be used.

Paradise Lost

onceived in 1959, Milton Keynes was the brainchild of Buckinghamshire's Chief Architect and Planning Officer, Fred Pooley. It was to straddle the West Coast main line and the A5, with the newly opened M1 to the east. Bletchley lay at the southeastern corner, with the Oxford–Bedford line running east–west a couple of miles to the south; at the northwestern corner lay Newport Pagnell, served by a branch line from Wolverton, on the West Coast main line. If ever there was an opportunity to build a peaceful new world with environmentally friendly public transport, Milton Keynes was it. Pooley had written of Milton Keynes that 'it is impossible at reasonable cost to plan a city satisfactorily for 100% motor car use' and that 'provision of public transport should be seriously investigated'. He proposed that a monorail would run east–west in long loops, through the new housing; people would pay for the transit system indirectly, as part

of their rates, meaning that no-one would be more than seven minutes' walk from 'free' public transport. Pooley's original plan did not envisage connecting the tramway to Fenny Stratford station on the Cambridge–Oxford line, but that could have come later.

Milton Keynes Development Corporation was formed in 1969 as a Government quango. Such Victorian hangovers as municipal electric tramways were not considered to be cutting-edge technology — which was a pity, because without new engineering projects the white heat of technology will cool down and the factories and employment along with it. The new men were in favour of the motor car, and acres of extra space were taken up to provide dual carriageways and roundabouts. Pooley's idea for a tramway may have had some influence, because the planners gave one or two of the dual-carriageway boulevards an extra wide, tree-planted

central reservation. In 2003 the suggestion is that electric trams could run along these from the outskirts to the railway — but only if the conservationists don't get up a petition to save the trees!

The queue of cars crawling through Bletchley to the railway station grew longer and slower, and the would-be railway passengers' prayers for deliverance rose skywards on the rather inappropriate incense of petrol fumes. Was this civilisation or what? British Rail had made up its mind in 1969 that 'for operational reasons' it was not possible to have a station at Milton Keynes! The shopping centre in Milton Keynes was opened in 1979, and still there was no railway station. By then road congestion had become serious and could only get worse, so British Rail began construction of a railway station for the huge city of Milton Keynes. It was brought into use in 1981 and was formally opened by the Prince of Wales on 14 May 1982.

The great blunder was that of not taking positive steps to encourage the use of railways as well as cars, of trusting that all transport needs could be solved by cars and roads: even I could see, in 1963, that this was a policy which carried the seed of its own doom within it. So surely the great brains of the quangos could realise that too? Perhaps they had been blinded by something. I am left with the nagging thought that this bias against railways and towards roads had everything to do with the interests of road-building contractors, car and lorry manufacturers and the oil companies rather than the peace of mind and wellbeing of the voters.

Left: Although Buckinghamshire's New Town was situated close to two existing stations — Bletchley and Wolverton — its centre was located at some distance from either of them. It was not until the early 1980s that the lack of co-ordination between town and railway planners was finally resolved with the opening of the new Milton Keynes Central station. On 24 March 1984 Class 87 No 87016 calls at the then relatively new station with the 11.05 express for Birmingham New Street. *John C. Baker*

Above: Milton Keynes was not the only New Town that had to wait some time for a new central station. Established in the 1960s, Telford possessed a number of stations — such as Wellington and Oakengates — that served the original communities united to form the New Town, but it had to wait until 12 May 1986 for the opening of Telford Central, which was built to serve its commercial and retail centre. This view, taken on the opening day, shows services heading Wolverhampton (left) and Shrewsbury (right). *Chris Morrison*

Lewes–Uckfield

The line from London to Lewes via South Croydon, Oxted and Uckfield was a vital main line in its own right, bringing trains from Newhaven and Eastbourne to London Victoria. It also constituted an effective quadrupling of the Brighton–London main line, which lay a few miles to the west: in the event of engineering work on the Brighton line, trains were diverted via Lewes and Uckfield. At Eridge, a little way north of Uckfield, a line branched out to Tunbridge Wells, giving Lewes people and the operating department of the railway another route to/from London, to enable lots of trains to be run to lots of destinations in London.

In April 1963 traffic congestion in Lewes was severe and East Sussex County Council (ESCC) was under pressure from the Lewes Chamber of Trade for a 'relief road' to take through traffic out of Lewes High Street. As a result it adopted a plan for a 'Lewes Relief Road', to be built in three stages, beginning with . . . the Phoenix Causeway — presumably rising from the ashes of the railway, as we shall see.

This first stage of the new road would have to cross — or cut through — the double-track main line from Lewes to London, and a bridge would cost £135,000.

In April 1963 the Beeching Report marked the Lewes–Uckfield line for closure to passenger trains — but not freight. How it could have been made profitable by withdrawing the passenger trains is one of the great mysteries of a report full of mysteries. However, the closure recommendation was a Godsend to the ESCC's Highways Department, because, with the railway out of the way, it could build its road through the railway embankment and save £135,000.

In spite of the Beeching recommendation and the enthusiasm of the Ministry of Transport for railway closure — and doubtless to the chagrin of the ESCC's Highways Department — the line was not then closed; someone, somewhere was trying to keep it open. In January 1965 one civil servant at the Ministry of Transport wrote to another: 'It now seems likely, according to the latest information from the [British Railways] Board, that publication [of the closure proposals for the railway] cannot be expected until the spring and therefore the earliest time the Minister's decision can be expected is early autumn — say September.' On 3 June 1965 another letter within the Ministry stated that 'No progress has been made regarding the statutory processes involved in closing the line and the Railway Board have now put to the County Council a proposal to re-open a section of closed railway known as the Hamsey Loop which would replace the present line and does not cross the line of the Lewes Relief Road.' The railway had been there for 100 years, the road did not exist but it was the railway that was crossing the road! The sentence illustrates well the mindset of the Ministry of (Road) Transport.

Closed in 1862, the 'Hamsey Loop' had formed part of the railway's original route into Lewes from Uckfield and had consisted of a 1,593yd stretch of line which joined the Lewes–Haywards Heath line ¼ mile west of Lewes station. By mid-1965 BR no longer owned the entire trackbed, and its lawyers were drafting a Bill to go to Parliament to give BR powers to reacquire the trackbed — and for the compulsory purchase of properties subsequently built upon it. BR estimated the cost of reopening to be £90,000 and suggested to the ESCC that, since reopening the Hamsey Loop would allow the existing embankment to be cut through by the road and thus save the ESCC £135,000, the ESCC might contribute towards this cost. The County Road Surveyor recommended this course of action in a letter to the County Council, stating that the County could expect any contribution made by the Council towards the cost of the Loop to be 75% grant-aided by the Ministry of Transport. Why could the Ministry not give BR a direct grant for 75% of the cost of the work? There was no provision for doing this — road schemes get grants, but BR pays 8% interest on its investments in new works and then gets blamed for making a loss. This seems to demonstrate the bias of governments against railways in favour of roads.

This unusually co-operative recommendation from the ESCC Highways Committee fell on stony ground — the railway was not to be treated as a friendly part of Britain but as some alien force to be suppressed wherever possible — and the ESCC refused to contribute anything to the cost of the Hamsey Loop, even though it would get

After the closure of the line south of Uckfield, the passenger services terminated at the original station in the town — as shown in this view of Class 207 DEMU No 1305 on 16 July 1975 awaiting departure for Victoria. More recently, the line has been shortened further, a new station being constructed to the north of the level crossing (effectively where the photographer is standing); whilst this may make sense operationally, it will make reopening the line more difficult, as there will no doubt be local opposition to the restoration of the level crossing and any consequent delays to road traffic. *E. W. Fellows*

back 75% of its expenditure. Its attitude was that, since BR would be saved the cost of maintaining three river bridges over the Ouse and the 350yd-long iron viaduct just outside the station, it was doing BR a favour by making it reopen the old line, so BR could pay the £90,000 reinstatement cost in full. Meanwhile the Phoenix Causeway would qualify for a grant from the Ministry of Transport.

While the Hamsey Loop Bill was being drafted, the road fiends were becoming agitated because their project was being delayed. East Sussex County Council had planned to have the Phoenix Causeway open by the end of 1966 at the latest. A letter from a civil servant at the Ministry of Transport to the ESCC Highways Department stated that 'if the Hamsey Loop Bill goes before Parliament, the expectation is that it will become law in July 1966 and that the new line would then be working by the end of 1967'. The letter concluded: 'I am afraid this news will be rather melancholy from your point of view. We for our part shall do our best to process the closure proposal as quickly as possible but I cannot promise any significant advance on the forecast decision date.' So all that mattered was their piffling bit of road, and a double-track main line to London was merely a nuisance.

In December 1965 Sir Tufton Beamish, MP for Lewes, was informed by the Ministry of Transport's civil servant that British Rail 'has not yet decided whether to go ahead with a proposal to withdraw passenger services' — which is an odd statement, given that BR was actually trying to reopen the Hamsey Loop to keep the line running and oblige the road lobby by getting out of their way.

The Bill for the Hamsey Loop became an Act in 1966, opening the way for reinstatement — but, amazingly, BR did not use its expensively won powers. The existing alignment was still in use in 1967 when David McKenna, General Manager of the Southern Region, stated publicly that the Lewes–Uckfield line was classed as a secondary main line and was included in his Master Plan — 'Blueprint for the Southern'. However, in spite of Mr McKenna's confidence in the future importance of the line and in spite of BR's having obtained an Act of Parliament to reopen the Hamsey Loop, Labour Minister of Transport Barbara Castle in 1968 consented to closure of the railway from Uckfield to Lewes with effect from 6 January 1969, subject to the granting of licences for a bus service to replace the trains — which buses would, of course, be subsidised from BR funds.

When the first meeting of South East Area Traffic Commissioners took place, on 6 January 1969, the Southdown bus company had not prepared its proposals, so the meeting was adjourned until 21 January and the trains continued to run. On the 21st the Southern Region Civil Engineer dropped a bombshell by announcing that the Lewes iron viaduct was unsafe; he would sanction its continued use — as a single track only — until 23 February. The bridge had been quite sound when Mr David McKenna spoke of the importance of the line but had apparently become terminally unsafe 18 months later.

From 6 January, therefore, an hourly shuttle service of trains began running between Lewes and Uckfield while the bus company was supposed to be organising extra bus schedules to replace the trains. On 23 February the train service was suspended because the viaduct was now said to be too dangerous to be used. Southdown had still not provided an alternative bus service, so the Southern Region was obliged to hire buses from Southdown to run to and from Lewes and Uckfield, calling at Barcombe Mills and Isfield. Passengers went to the railway station booking office of their choice to buy railway tickets for use on the bus. Because these were double-deck buses they were unable to get from the A26 to Barcombe Mills station. To convey passengers between the station and the bus stop on the main road, British Rail hired a minibus. So passengers either walked or drove their car to Barcombe Mills, bought a ticket and boarded the minibus for the mile back to the bus stop on the main road. During this period the River Ouse flooded and the minibus service was suspended. The BR bus service continued until 4 May, when it was taken over by Southdown.

One month exactly after closure of the Lewes–Uckfield line, a derailment south of Balcombe Tunnel on the London–Brighton main line stopped all trains and caused severe delays and inconvenience to thousands of passengers. Later in the year further derailments and landslides closed the Brighton line, with similar consequences for passengers. Unreliable train services bleed passengers to the roads, which is good news for the petrol companies but bad news for people — and for BR's income, which continued to decline.

To truncate the line at Uckfield made absolutely no sense, financially or from an operating point of view. It became a mere branch line with higher operating costs, no major destinations served, fewer passengers to use it and with no operating advantage as a diversionary route. British Rail was saddled with the cost of maintaining the dead-end 'Uckfield branch' rather than a secondary main line to Brighton, Eastbourne and Newhaven, and the increased loss to BR finances would doubtless be trumpeted in newspaper headlines the following year.

In December 1969 6,810 people were killed on the roads of England and Wales. The total cost of road accidents in 1986 was estimated by the Ministry of Transport — which shuts down railways — at £3.8 billion; urban roads like the 'Phoenix Causeway' accounted for 76% of all road accidents, at an average cost to the nation of £12,560 per accident. None of these figures was admissible as evidence for the benefits of having a railway line. The Polegate bypass, less than 4 miles long, cost £30 million, but £25-29 million is too much to spend on re-laying the track the 7 miles from Uckfield to Lewes.

In 1975 East Sussex County Council repented slightly its earlier madness by giving 'protected' status to 7 miles of trackbed, including the Hamsey Loop. And there it lies to this day, as lineside populations grow, car use and inevitable congestion spiral, and while those in power still talk in outdated terms of 'profit' for the line instead of looking at what the line could do to take congestion and death off the roads.

If the route had remained a through main line it would have been electrified in 1980. Because of its declining receipts, BR reduced expenditure on its maintenance until, in 1985, the branch to Uckfield was on the verge of closure from sheer neglect. Severe speed restrictions were imposed in 1987, and in 1988 it was converted to single track with long passing loops. Still it was neglected, and the rolling stock was in poor condition. Grimy signal glasses and non-functioning AWS equipment on a train contributed to the terrible head-on collision between two trains at Cowden on 15 October 1994. The decayed aspect of the line, the slowness, the rough ride and the reduced train service ensured that fewer and fewer people used it and that consequently BR's losses on it increased.

And what of the fatuously named Phoenix Causeway? The congestion on this road quickly became so bad that even East Sussex County Council could see there was no point in continuing with Phases 2 and 3. The full scheme was never completed, and Phase 1, which brought about the closure of a major railway, was diverted southward to join the A27 Lewes bypass. This required the construction of the Cuilfail Tunnel — no expense spared, unlike railways, which are to this day 'too expensive' to reopen.

After closure, a preservation scheme — the Lavender Line — developed at Isfield station, which was lovingly restored. Although the restored line stretches north of the station for a short distance, as evinced from the buffer-stops, Isfield is the southern terminus. Proposals for the reopening of the line between Uckfield and Lewes rumble on, but, unless somebody is prepared to take the project forward in a determined way, these are likely to end as yet another post-privatisation pipe-dream.
Ian Allan Library

Haverhill

Between 1960 and 1965 the little Suffolk town of Haverhill, standing in the vastness of the most rural countryside and 'miles from anywhere' (although it could be said to lie between Cambridge and Colchester), was being expanded. Thousands of new houses were being built, to accommodate displaced populations of working folk from London. The roads were delightfully 'olde worlde', and buses found the route northwest to Cambridge all but impassable. People had voluntarily moved out of East London to this wonderful area believing that they would have a railway to get them to their work, as the Cambridge–Sudbury–Marks Tey line passed through the town. However, in 1967 the Government of the People closed the railway between Haverhill, Sudbury and Cambridge to save money and spent it on opening out the roads, thus doing its bit for the road-building contractors and traffic congestion while increasing the costs of the National Health, Police and Fire services.

Pictured a month before closure, in February 1967, this was the nameboard outside the station. As with countless thousand other stations closed in the 1950s and 1960s — and, indeed, later — part of the community's social fabric was lost with the demise of the railway, and life for the young, the elderly and the handicapped suddenly got much more difficult. *G. R. Mortimer*

On Saturday 4 March 1967 — the last day of passenger trains between Cambridge and Sudbury — passengers wait to board the 10.55 service from Cambridge; from the following Monday — when passenger services were officially withdrawn — they would have no choice but to make their journey by car or bus. The rump of the line, from Mark's Tey to Sudbury, was also slated for closure on several occasions but has survived to the 21st century, ensuring that at least part of this growing area retains some access to the railway network *G. R. Mortimer*

With the Benefit of Hindsight

The history of railways in Britain is littered both with examples of decisions made that, after the event, can be regarded as unfortunate and with opportunities that, had they been grasped at the time, might have ensured a more efficient or better railway network. Undoubtedly, hindsight is a wonderful attribute, and this section highlights a number — by no means exhaustive! — of these decisions and opportunities.

During the 1960s, a number of lines were closed that, with the benefit of hindsight, perhaps should not have been dispensed with. Amongst these was the ex-Midland Railway main line from Matlock to Buxton and Chapel-en-le-Frith. In the view on the right, Class 4P No 41185 heads into Matlock with a service from Manchester. Following closure, the line was dismantled, but since the late 1970s Peak Rail has been endeavouring to reopen the line and currently operates from Matlock (Riverside) to Rowsley (South). One of the PR's problems is that bridges across the A6 and at Buxton have been demolished. More recently, the SRA has suggested that reopening of the route could form part of its future investment plans, although there is no timescale for this. *D. Sellman*

Initially it looked as though the construction of the North British Waverley route from Hawick to Carlisle was a blunder in the 19th century, but traffic eventually developed to justify its retention. However, in the 1960s, with various competing Anglo-Scottish routes, the line from Carlisle to Edinburgh via Hawick was deemed surplus to requirements, despite the huge investment in the Millerhill marshalling yard. The line closed in January 1969 amidst much opposition but refused to die. In 2002 passenger services were restored over part of the northernmost section, and there are proposals for reopening further sections at both northern and southern ends; however, as is often the case, piecemeal demolition (notably of the viaduct at Hawick) or reuse (such as the conversion of the trackbed at Melrose into a road) makes reopening more complicated or expensive. Here, on 18 August 1966, Brush Type 4 No D1968 heads south with a freight at Melrose. *C. Lofthus*

Above: Few of the Victorian entrepreneurs had the perception of Sir Edward Watkin. He had a grand vision of the construction of a main line from the north of England to Europe via a Channel Tunnel. Granted, he was a century ahead of his time as far as the tunnel was concerned, but in the Great Central main line he produced Britain's only major route constructed to the Continental (UIC) loading gauge. At the time, the Great Central Railway was mocked as the 'Gone Completely', and it is unlikely that, during its 70-year life, it ever fulfilled the expectations of its promoters, but it did represent a fast and direct link to the industrial Midlands and North prior to its closure. Today a private consortium, Central Railways, is seeking to restore the line but is confronted with the fact that much of the infrastructure, such as the viaduct at Brackley (shown here on 27 March 1965, with a Class 5 4-6-0 crossing with the 2.30pm service to Nottingham), has been demolished. The frustrating thing about this particular viaduct was that it survived for more than 20 years after closure before being blown up just as the Channel Tunnel project was being more fully developed. *Brian Stephenson*

Above: There were two competing lines from Exeter: the ex-Great Western line via Dawlish and the ex-LSWR line via Okehampton. In the late 1960s and early 1970s, when rationalisation was the 'buzz word' and elimination of duplicate lines the creed, it was decided to concentrate resources on the ex-GWR route, and the line via Okehampton was abandoned west of Meldon Quarry. This was fine, provided that the ex-GW route was guaranteed to be open permanently, but the Dawlish sea wall has proved to be problematic and increasingly prone to closure, thus severing west Devon and Cornwall from the rest of the railway network. Whilst thought has been given to reopening the ex-LSWR line, this is probably now impractical. Here, on 1 March 1979, Class 47 No 47479 heads towards Meldon Quarry at North Tawton over the surviving section of the line. *Peter Medley*

During the early decades of the 20th century the Great Western Railway constructed a number of cut-off routes, providing better connections between major cities and additional capacity. One such route was the line between Stratford and Cheltenham via Honeybourne. Whilst passenger services ceased in the 1960s, freight continued to use this important diversionary route until the early 1970s, when, following a derailment, the line was closed. By this time it provided the only alternative route between the Midlands and the South West, so its loss was unfortunate — the more so as the track was not lifted until almost a decade later, by which time an active preservation scheme was endeavouring to save the route throughout. Preservationists have now restored a link between Cheltenham racecourse — seen here in January 1966 — and Toddington, but this is another route which Railtrack had at one stage identified as having potential for reopening. *Andrew Muckley*

In the late 1950s considerable investment went into the quadrupling of the East Coast main line to the south of Welwyn Garden City, with second tunnels being excavated at Hadley Wood and Potters Bar. The only exception to this major investment was the section to the north of Welwyn over the viaduct and through Welwyn North station. Although there were undoubtedly sound reasons why this work was not undertaken at the time (most notably the difficulty of constructing a second viaduct), the failure to address the problem in the 1950s has resulted in the two-track formation between Stevenage and Welwyn Garden City becoming a significant restraint upon the future development of the line, which over the past decade has witnessed a significant increase in traffic volumes. This view shows Welwyn North station prior to rebuilding but more than 40 years after the quadrupling south of Welwyn. Despite hopes to the contrary, prospects for improving the section through this station remain forlorn. *Ian Allan Library*

It is often said of Britain's railway network that when Birmingham New Street sneezes the rest of the system gets a cold. Whilst this may be a slight exaggeration, it does hold water for the complex Virgin West Coast and Cross-Country franchises. One of the fundamental problems with the station is that, at the southern end, there are only four roads through the tunnel on the line towards Coventry. Towards the end of the 1990s there was the opportunity, given the reconstruction of the Bull Ring Centre, to have constructed an additional two lines southwards; however, the opportunity to increase flexibility — by widening the approaches to the station throat — was lost, condemning New Street to a further generation of problems. Here, in September 1970, slow and fast trains await departure from New Street on services to Euston. *M. Dunnett*

In the late 1930s it was proposed that the ex-GNR lines serving Edgware and Alexandra Palace be transferred from the LNER to the LPTB and electrified as part of an expanded Northern Line. Work was undertaken, including the erection of trackside supports for cabling over the route to Alexandra Palace, and the part of the network does indeed survive as part of the LUL system. Unfortunately, World War 2 and postwar austerity intervened to ensure that the proposals were never completed. In July 1954 the remaining passenger services from Finsbury Park to Alexandra Palace were withdrawn. Fifty years on, despite well-publicised problems, Alexandra Palace remains a popular destination and exhibition centre, and, every now and again, proposals for the reopening of the line resurface. This view shows well the close relationship between the GNR-built station of 1911 and the palace. *J. N. Young*

The
Isle of Wight

The Isle of Wight's steam railways from Ryde Pier to Ventnor and Cowes were wonderfully effective. Those who had an interest in stealing their traffic could point to the 'extreme age' of the locomotives and stock, but what they did not like to comment on was the astonishing efficiency with which these trusted machines moved tens of thousands of people. The working railwaymen of the Island did a wonderful job; it was the Government which lost the will to keep the railway running.

The motive power and coaches were allowed to become run-down, and in July 1963 four of the 0-4-4Ts were withdrawn after their cylinders were found to be dangerously thin-walled. In the same month another of them failed on an 11.30 Ryde–Cowes train. With 20% of the locomotive fleet unserviceable, the 1963 summer service could not be implemented. Some of the relatively new BR Standard '84xxx' tank engines could have been brought over from the mainland and given a useful life extension on what was still a busy railway; there would have been a clearance problem at Rink Road bridge, but surely the bridge could have been rebuilt? Some of the existing locomotives could have had liners pressed into their cylinders to bring them up to standard, whilst the very comfortable Victorian and Edwardian carriages could have been refurbished; the Island was a special place and required special treatment. The idea was discussed and abandoned.

Ryde Pier Head station was closed for the winter of 1963/4. Passengers getting off the ferry at Ryde Pier had to walk to Ryde Esplanade if the pier tram was full, which it would be after taking on a few dozen people. The timetable had an 'N' in all the columns, but nowhere was this note explained. What it meant was that all trains from Ryde Esplanade were for the Ventnor line, passengers for

On 20 March 1967 a sorry line-up of steam locomotives awaits its fate at Newport station, following the closure of much of the Island's railway network. As is apparent, Newport was provided with extensive facilities and was one of the Island's major centres to be deprived of railway services. *Alastair McIntyre*

Cowes being expected to change at St Johns. Newport-line passengers would (if they could hear them) discover such important facts from station announcements — otherwise they would find themselves in Brading.

That great industrialist, Dr Beeching, saw nothing of use in the Island's industrious but rural railways. No fewer than 2,800,000 people a year were then using the trains — indeed, it was not uncommon for 1,000 people to come off a single ferry — yet these passengers were to be subjected to the closure treatment.

On 20 April 1964 the Islanders were informed that their railways would be closed in October. In 1955 BR had given the Islanders an undertaking that they would be given five years' notice of closure for the Cowes line and seven for the Ventnor, but BR felt absolved from this promise because the lines were recommended for closure in the Beeching Report.

The Islanders made a fight for their railway. BR thought that the line between Ryde, Newport and Cowes might be retained, since there was a steadier number of people using it all year round, but Barbara Castle took the decision to allow this route to be shut and the holiday route to Shanklin retained. But in Parliament the real reason for the attack on the Island's railways became clear: it was stated that the Island's roads were long overdue for widening and straightening, and there was a need for bypasses. There was not then, but there

would be if the railways were closed. The roadbuilders were not getting the contracts that they would if the lines were shut. All the horrors of the tarmac serpent were to be inflicted upon one of the few idyllic places left in the South of England — people went there to enjoy wonderful villages, relatively traffic-free, served by a well-run railway.

It would be too much to expect any imagination to be used to supply the idyllic Island with something dignified. In 1967 BR imported from the mainland redundant London Underground tube trains dating from 1923-35; Ryde Tunnel was liable to flooding, and, while this was not a problem for steam engines, it would be for third-rail electrics, hence the floor of the tunnel was raised, making it impossible for anything other than a tube train to pass through.

Did the people of the Isle of Wight get a 'world class' railway after privatisation in 1994? Did private enterprise do any better than British Railways? That great herald of free enterprise, Stagecoach, took over the 'Island Line'

When the decision was made to electrify part of the Island's network, it was also decided that trains should thereafter terminate at Shanklin, with all services thence to Ventnor abandoned. On 19 August 1981 the 14.41 service for Ryde, formed of units 043 and 033, awaits departure. In the early 1990s these units were replaced by newer ex-LT stock, albeit still pre-World War 2. Today, some 40 years after its withdrawal, there is pressure for restoration of the link to Ventnor. *Jeff Dacombe*

franchise — the Ryde Pier–Shanklin line. Even with electric trains, a skeleton track layout and daily and seasonal peaks the line cannot be made to pay — even with the dazzling talent of private enterprise. In 1996/7 Stagecoach took a subsidy from the taxpayer of £2,100,000 for this franchise. If BR were running this, then at least the subsidy would be less because it would not have to be shared amongst Stagecoach shareholders. To cut costs Stagecoach singled the double track between Brading and Sandown, leaving Sandown as the crossing place. The normal service is two trains an hour leaving Ryde Pierhead at 20 and 40 minutes past the hour. The Wightlink ferry from the mainland is no longer run by the people who own the railway and arrives at the pier at 30 minutes past the hour. The xx.20 train leaves as the catamaran is in sight, and one hopes that the xx.40 can hang around for a extra few minutes to allow the crush of people to leave the vessel and board the train. In the high summer season, from late July to early September, the Wightlink company pays Stagecoach to run an extra train to give a 20min frequency from Ryde.

'Liner' Trains

In 1963 Dr Beeching's report identified 15 routes for 75mph 'Freight Liner' trains, to carry containerised freight and parcels. Initially there would be trains linking London, Liverpool, Manchester and Glasgow; these services were to be running by late 1964 or early 1965. Brand-new terminals were built at Manchester (Longsight), Liverpool (Garston), Glasgow (Gushetfaulds) and London (Maiden Lane), costing BR £320,000. The trains — themselves an entirely new concept — were built at Ashford Works, and the total investment made by the BR Board in this road-beating idea was £6 million. (A curious aside: Conservative Minister of Transport Ernest Marples refused to allow the railway workshops to build the containers that would ride on the trains and the privately owned lorries. This weird piece of prejudice was a severe blow to the economy of the railway workshops and caused resentment among the unions' leadership. Tom Fraser, Minister of Transport in the 1964 Labour Government, duly rescinded the Marples edict.) Dr Beeching and the BRB confidently expected that 40 million tons of traffic could be taken off the roads (and that was just a start) as the benefits of the system became apparent and more depots and routes were opened.

The scheme was dependent on any factory or road haulage company being allowed to drive its ready-loaded freight in containers to the new railheads, but this the National Union of Railwaymen would not permit. The NUR had earlier criticised the Beeching Report for looking at railways narrowly without realising the additional costs that would be imposed on the public purse through road building and road accidents. Now the Union took the narrow view: allowing private hauliers into railway depots would take away the work of its members in the railway cartage and loading division and hand it to outsiders — many of whom were members of the Transport & General Workers' Union. Without the modernised Freightliner trains BR's freight carrying would die, but the Union was adamant in its ban on outside hauliers, and three years of non-development — three years of lost revenue — ensued.

In March 1967 Labour Minister of Transport Barbara Castle threatened to withhold all investment in the railways if the NUR did not withdraw its ban, and the

One of the new container terminals was at Gushetfaulds, near Glasgow, as seen in this contemporary publicity photograph with Class 47 No D1634. Unfortunately, whilst the container was perceived to be the future for rail-borne freight, the railway unions were not as convinced as the management, with the result that services would not be developed fully until the late 1960s. *British Rail*

Union gave in; it assumed that in so doing it would be acting for the greater good of the railway industry. The following year, however, the 1968 Transport Act removed this profitable business from the railways' budget and transferred it to a new, state-owned business — the National Freight Corporation. Sir Philip Warter, recently Dr Beeching's assistant, was appointed Chairman of the new enterprise. Warter it was who in 1964 (when he was in charge of British Road Services) had said that it was his job to run a profitable freight company and that he did not, therefore, see the point in putting containers on rail from London to Birmingham or Manchester when there was a perfectly good road. This underlines the difference between a narrow-minded private-enterprise approach and the wider view of the cost benefits to be gained by keeping heavy freight off the roads.

Single-lead Junctions

The single lead is a cheap method of laying out a junction and saves the costs of materials and maintenance associated with a conventional (but more complex) double-line junction. However, it reduces the number of trains that can pass through the junction and introduces what did not previously exist at a double-track junction — the possibility of a head-on collision if a driver over-runs the junction signal at Danger. The new form of junction was invented by British Rail in 1958 and was permitted by the Chief Inspecting Officer of Railways because it was cheaper than that which went before — not because it was safer, which it wasn't. British Rail stated that 'The single-lead junction is easier and cheaper to maintain and therefore it is less prone to dangerous

A traditional — if complex — junction layout: Lansdown Junction, Cheltenham, in late 1945, showing the work involved in route-widening at this point. *GWR*

deterioration than the conventional junction'. This reasoning implies that, in the past, gangers who found a junction difficult to maintain safely just gave up and let it go! The Railway Inspectorate and British Rail also asserted that 'The single-lead junction is no different in principle from countless other situations where trains run in both directions over a single track'. That sentence could be re-written with more truth: 'The single-lead junction is as dangerous as countless other situations where trains run in both directions over a single track.' And, of course, by substituting a single-lead junction where there had been a double-line junction, the number of potentially dangerous situations was increased. The single-lead junction brought about three head-on collisions in 2½ years. After the third accident a moratorium was placed on the installation of single-lead junctions, and three high-powered risk-assessment inquiries were conducted into the relative dangers of the single-lead junction and

the conventional double-line junction. (One is bound to ask whether such a risk assessment was carried out in 1958.) All three studies reached the same conclusion as would any ordinary railwayman — that there was a greater risk of a collision at the single-lead than at the double-line junction. Three train drivers and three passengers lost their lives, and 50 were injured, as a result of this outrageous obsession with cutting costs until the railway is reduced to something which can only be described as 'cheap and nasty'.

Although no more single-lead junctions are to be installed, instances of these dangerous layouts being reinstated as double-line junctions are at best rare. The most likely reason for reverting to a conventional junction is not safety but commercial. Given a situation of greatly increased traffic, the delays likely to be caused by having only a single line through a junction will become intolerable, and layouts will be rebuilt according to more efficient — and safer — 19th-century principles.

Above: Work is also in progress at Marshgate Junction, north of Doncaster, in this view taken in March 1979. To the right are the lines heading northeastwards to Goole and Hull, reached via a ladder junction. *British Railways*

Right: The single-lead junction at Hyde North station, with (in the distance) the slip points introduced to protect the junction following a collision at this location on 22 August 1990. *Stanley Hall*

Class 313

In 1968 a prototype EMU (Class 445) was designed by a non-railway drawing office. Following two years' trials on the Southern Region — which were deemed to have been a success — the design went into full production, but these units differed from the prototypes in terms of major components. This 'new' design was not subjected to any service trials but as Class 313 was put into revenue-earning service in 1976 on the newly electrified GN-line suburban routes. By 1981 the class had achieved an impressive record, running an average per unit of 7,600 miles between failures. In 1980 the antique Class 306s, designed by railwaymen and put into service in 1949, were running 19,940 miles between failures — a record 250% better than that of the brand-new machines; the Class 309s, in service since 1962, were managing 28,800 miles between failures. In March 1981 Secretary of State for Transport Norman Fowler praised the contribution privately owned manufacturers of railway rolling stock were making to the railways and promised to give BR money to spend with these private companies. What was it that he was praising?

The major problems on the Class 313s concerned the sliding doors and automatic couplers, followed closely by the air-compressors and tripcock gear.

The air-operated mechanisms of the sliding doors were exceedingly complicated and sensitive to a speck of dirt, which for railway operation is not sensible.

The couplers were supposed to permit the three-car sets to be coupled into longer sets at peak periods. Unfortunately the prototypes were equipped with German couplers, totally brilliant in Germany but entirely useless in Britain. So an adaptor to go on top of the German coupler was chosen for production models, an American 'Tightlock' coupler. This was for emergency purposes only, however, and a new coupler for daily use had to be designed. British Rail again made the mistake of asking an outside contractor to design and supply the new couplers, and that firm promptly sub-contracted the job to another firm. When two firms are both trying to get their profit out of one price there are bound to be difficulties later. The new couplers had poor weather sealing, allowing water into the electrics, were of poor design, allowing sealing rings to fall out, and featured a badly designed locating probe, so that coupling could only take place when two sets were on a straight section of track. As a result BR had to keep the units in six-car formations all day rather than risk being unable to couple sets at peak times. Quite apart from the costs involved in paying contractors for shoddy work, the cost of the extra electricity required to run six-car trains when three cars were all that was needed amounted to £700,000 per annum.

The air-compressors were to supply air for the brakes as well as for raising the pantograph and operating the doors, suspension and traction control. The problem was that in 1976 an 'off the shelf' compressor that was powerful enough to do the work was not available. Rather than design and build one suitable for arduous railway service, the contractors, with BR's agreement, installed a ready-made compressor that was too small for the work to be done. The nonsense was compounded because the chosen compressor had basic defects in its design which caused it to fail. By 1980 a more powerful compressor had become available, but BR baulked at the high cost of replacement.

The tripcock-activating device on the Class 313s was needed when they were used on LT — Drayton Park to Moorgate. The original equipment was too flimsy to withstand the blow it encountered from the cast iron of the LT lineside tripcock, and between 1976 and 1979 all 64 units had had their tripcock-engaging levers replaced twice! It all goes to show that the proper people to design and build railway equipment are railwaymen.

A posed publicity shot for the Class 313 EMUs when they were first introduced in 1976. As described in the text, these units had countless teething problems, but, almost 30 years on, the type continues to give sterling service on suburban services out of King's Cross and on Silverlink Metro services. *BRB*

Retreat from Traffic

The retreat from traffic was always a matter of frustration for railwaymen — especially when they saw the trouble taken over road building. It is not too much trouble to employ vast earthworks to create flying junctions and slip roads at Stansted or to widen the M25, but a relatively little matter like relaying the track from Cambridge to Sandy to create an absolutely vital east–west railway is simply too difficult and will never be done. This deathly notion of profit came to infect the nationalised railway. It was a Labour Government in 1968 that charged it with making a profit on its operations rather than looking at the wider picture of rendering a public service to the nation — 'keeping death off the road', as the old boys used to tell me. Milk trains were abandoned when new tankers were needed. Outer-suburban branch lines around huge cities were closed. They were said not to pay — but closing them created an enormously expensive problem of transport into the conurbations. 'That's not an expense on my balance sheet,' said the one-time General Manager of Western Region, Sid Newey.

The Government of the 1980s withdrew grants to private sidings, with the result that they were closed and the traffic went onto the roads. The maltings at Diss and Great Ryburgh had their sidings from the main line closed, whereupon 38-ton lorries began to force their way through narrow lanes. Branch lines serving road stone quarries were, of course, not shut — they *must* be profitable. Under nationalisation and privatisation the railway has been curtailed and strait-jacketed, even though past policies have ensured that the roads are so congested that people are now clamouring to travel by train. Some bits of lost line are now being brought back into use, timidly and at great cost, notably the Robin Hood-line extension and the reinstatement of the double track between Princes Risborough and Aynho Junction, the latter an epic of determination by Chiltern Trains in the face of stiff resistance from the Strategic Rail Authority.

It is always a struggle to increase rail use, but rail land is easily sold off. Evesham yard, which might have been used an 'inland port' by freight operator Christian Salvesen, was instead sold off for redevelopment by a

Milk wagons were highly specialist, both in their construction and in their use, being formed of a glass container within the metal body. For countless years the railway provided a safe and efficient means of shipping large quantities of milk into London, using wagons like these six-wheel wagons built for the GWR and seen at Swindon in December 1980. However, despite the location of many dairies close to the railway line, this traffic was lost, and recent efforts to return milk traffic to the railway have failed. The population of London still consumes vast quantities of milk; unfortunately, it is now all delivered by road.
G. Scott-Lowe

Above: The Chiltern line north of Bicester was singled and, indeed, was slated for further rationalisation, including the removal of the Aynho flyover. Moreover, most through services thereafter ran only between Banbury and London, the traditional link between the route and the West Midlands effectively being severed. Fortunately, wiser counsel prevailed, and many of the more radical reductions never occurred. Such has been the growth of traffic over the line that double track has now been reinstated throughout, and Chiltern Railways provides a useful link between London and Birmingham, restoring the traditional GWR/WR route to the north. In the dark days, on Sunday 12 April 1981, the 14.30 from Marylebone to Banbury departs from Bicester and prepares to enter the single-track section. *John Acton*

Below: Another route to suffer seriously from singling was the ex-LSWR main line between Salisbury and Exeter, as evinced by this view of Class 50 No 50015 approaching Tisbury on 19 October 1986 with the 09.40 service from Exeter to Waterloo. With the significant increase in traffic over recent years, there has been partial redoubling, but there are still long single-track sections, which can cause serious delays if a train misses its path. *W. A. Sharman*

Above: One of the freight-only lines that disappeared relatively late was the route from Boscarne Junction to Wenford Bridge, which survived until the early 1980s for the movement of china clay. Unfortunately, modernisation of the equipment for the china-clay traffic resulted in the new rolling stock being too large for the lightly engineered line and thus it closed. The route from Boscarne Junction to Bodmin Parkway was preserved, and, in order to reduce the number of lorries running over the narrow North Devon roads, consideration was given to reopening the route to Wenford Bridge. Unfortunately, however, the trackbed had been turned over to a cycleway, and it proved impossible to merge the interests of both cyclists and railway. Here, on 8 July 1960, one of the diminutive 2-4-0Ts for which the line was famous, No 30587 (later preserved), heads towards Wenford Bridge at Dunmere Junction with the daily freight from Wadebridge. *J. C. Haydon*

Cornwall

Lower right: For more than 160 years the railway industry has been inextricably linked with the Royal Mail. Travelling Post Offices traversed the country, carrying the cheque and the Postal Order — as well as providing countless other services. With the election of the Labour Government, with its desire to see more freight carried by rail, one might have assumed that the Royal Mail would be encouraged to develop its relationship with the railway; in the event, despite huge investment in new terminals (notably at Bristol and Stonebridge Park) and rolling stock, the Royal Mail announced in mid-2003 that by mid-2004, as a result of problems post-Hatfield, it would cease using rail as a means of transporting mail. Here, on 7 April 1998, one of the specially constructed postal EMUs, No 325009, passes Shortlands with a service from Willesden RDC to Tonbridge. Less than a decade old, these units will presumably have no work once all the mail is transferred to road or air transport. *Brian Morrison*

supermarket chain. In 1965 BR revised its costs of shunting within docks and raised its charges 'beyond the capacity of any industry to absorb'; as a result Associated British Ports ceased using King's Lynn goods yard, and in 2000 Railtrack sold the site for redevelopment as . . . a supermarket. Leeds carriage sidings, meanwhile, have been turned into a shopping mall, and the site of Bristol Bath Road depot is also to be sold off for redevelopment. Any spare land is lost, so that there is no room to expand the railway.

Where platforms are long enough for the trains they are shortened, where they are not long enough it is very difficult to get them lengthened. At Cark & Cartmel (First North Western) the platforms were able to accommodate two three-car sets of Class 156, with room to spare. They were shortened during 2003 so that such a train overhangs by one door at each end. What is the point? The area is a magnet for tourists, and in summer two three-car sets are a necessity. First North Western is actually going to introduce Class 175 DMUs to the line and knew that when it shortened the platforms. But the Class 175s are longer than the Class 156s. First North Western wants drivers to open doors and to dispense with guards. But if the train overhangs the platform by several

Above: Wagon-load freight is dead,
long live wagon-load freight.
Although the post-Beeching era saw
an inexorable retreat from traditional
wagon-load freight operations, the
railway industry continued to seek a
means of attracting this type of traffic.
One such initiative was Speedlink,
developed in the late 1970s and
early 1980s. Here Class 37 No 37412
arrives at Crianlarich with the
Corpach–Mossend Speedlink service
on 4 June 1986. Despite high subsidies,
Speedlink failed to survive, but
another initiative — Enterprise —
by Transrail, one of the three freight
companies established prior to
privatisation, sought to reinvigorate
the business. This too came to nought
following privatisation, when the new
private-sector owners pulled the plug.
W. A. Sharman

doors, what then? Why not just leave the platforms alone and take advantage of that to run long, strong trains with lots of holidaymakers and a happy traincrew.

On the slow lines between Northampton and Euston, Silverlink is going to run 12 coaches instead of eight. There are lots more passengers these days, and the extra seating capacity is badly needed. At Bushey, just south of Watford, a thousand people per day use the station, all commuting into London. Silverlink stops at Bushey at peak times and whisks the crowds away at high speed — Harrow and Euston only. But when the 12-coach trains come into service they will not stop at Bushey — as things stand at present — because they are too long for the platform. Two coach lengths need to be added. The room is there at the south end. A simple job — the work of a weekend. But it has been the subject of months of meetings and nail-biting as to the cost. Just outside Bushey station the road traffic is at walking speed for most of the day.

The Paddington–Princes Risborough–Bicester–Birmingham service was downgraded early in 1967 after the inauguration of electrified Euston–Birmingham services. In 1968 the double track between Princes Risborough, Bicester and Aynho Junction was singled.

The really strange thing about these planners was that they knew that they wanted 'growth' in the British economy and yet it never occurred to them that they were throwing away the capacity to cater for that growth. Of course the greatest growth was in road traffic, yet, in spite of the M1 and M6 (and as I foresaw in 1963), roads alone could not cope. In 1998 a start was made on restoring the Princes Risborough–Aynho route to double track. At this time the cost of replacing 18 miles of track from Princes Risborough to Bicester was £30 million; extending the double track to Aynho Junction, where the line joins

the Oxford–Banbury route, cost, in 2002, £60 million for nine miles of track.

Honeybourne station was closed in 1969, when the village and surrounding area was growing in population. BR said it could not afford a station and hoped the county council would provide one. After the immediate vicinity had doubled in population the station was reopened on 22 May 1981. The new station consisted of a single concrete platform, the length of a three-car DMU, alongside the single line which was once a four-track station. On the platform was a plastic waiting-shelter and a 'portaloo'. It was opened to the sound of a brass band (which was nice), but once the bandsmen and councillors had departed the portaloo was vandalised, the waiting-shelter covered with graffiti — not the railway's fault, but a more positive approach with a proper building and staff might have been more encouraging for crowds of people to use. Unstaffed stations must be a real disincentive to trains, particularly after dark. And so the culture of yobbishness is given somewhere to practise its dark arts.

The Victorians believed that railways represented the height of civilisation. There is a strong tendency towards a retreat from civilisation in our free-market, individualistic world, and the retreat from railways is just a part of that movement. As I said in 1965, 'End of Steam — End of Civilisation'.

The somewhat basic facilities provided by the new station at Honeybourne are all too evident in this view of passengers waiting to board the 09.05 Oxford–Hereford service on 3 June 1982. The Oxford–Hereford route was another of the lines to undergo severe rationalisation in the 1970s; indeed, at one stage there was a serious possibility that it would close completely. *John Glover*

Roads Preferred

In 1988 work began on the Newark Relief Road (A46) as part of a £12 billion programme of new building. Running alongside the Nottingham–Lincoln line, this crosses the River Trent twice and bridges the East Coast main line 200yd south of the point where the Nottingham–Lincoln railway line crossed it on the level. The Newark railway level crossing has always been an obstacle on the ECML: conflicting movements cause delays, while the crossing itself is obviously more expensive to maintain than plain track. With the road running parallel to the Nottingham–Lincoln line, the obvious thing to do was to carry the railway with the road over the ECML. But, of course, that would have been attended with insuperable problems — quite apart from the awful cost of it.

Above right: The East Coast main line between Peterborough and Doncaster was constructed over relatively flat terrain, as a result of which, at two locations, it would ultimately be crossed on the level by other lines. As speeds increased, so these level crossings became increasingly restrictive operationally. At Retford, in the late 1960s, the railway funded the construction of a new underbridge with low-level station platforms. Here, on 21 July 1970, Class 37 No 6962 arrives at Retford with the 07.22 Harwich–Manchester service; on departure, it will pass under the East Coast main line. *David Wharton*

Right: The contrast at Newark could be no greater. This view, taken at Newark New Junction on 6 March 1965, shows Brush Type 4 No D1563 heading south with the 13.45 Doncaster–King's Cross express over the level crossing with the ex-Midland line between Nottingham and Lincoln. Today, whilst the railway infrastructure is effectively the same, the view would be dominated by the presence of the main A46 passing over the railway lines. Possibly, with a bit of pre-planning or judicious use of budgets, this railway bottleneck could have been obviated at the same time. Unfortunately, however, transport co-ordination is not something for which governments are conspicuously famous. *J. N. Smith*

The Ones that Got Away

By the mid-1930s the LNER was already establishing an impressive collection of preserved locomotives — the core of the future National Railway Museum — and when the ex-North British Railway Atlantics were due for withdrawal, the decision was taken to retain NBR No 875 for preservation as the last of its class. Unfortunately, the instruction to preserve the locomotive reached Cowlairs Works after the locomotive had been broken up. Nothing daunted, the works rebuilt No 875 from the surviving parts of the original locomotive and its sisters, then restoring it to main-line operation. As such, it soldiered on into World War 2, when it was again withdrawn. In the fog of war, when scrap metal was a vital resource, the locomotive was again scrapped, but this time there would be no reprieve. *Ian Allan Library*

Introduced to a design of Dugald Drummond in 1898, the Highland Railway 'Small Bens' were amongst the company's most versatile locomotives. By the early 1950s, only two remained in stock, and when the last, No 54398 *Ben Alder* — pictured here at Georgemas Junction on 5 July 1951, waiting to shunt stock the through coaches from Thurso on to the rear of the 9.45am Wick-Inverness service — was withdrawn, there were hopes that it would be preserved. Indeed, it spent the next 14 years anticipating such a fate but was scrapped in 1967. The reason? Like most of its class, it had been rebuilt during its career and was not, therefore, 100% original. The victory of the purists cost preservation one of the few ex-Highland locomotives to be a potential preservation candidate. *Ian Allan Library*

It was not just locomotives that failed to make it into the modern age; a number of other preservation schemes also failed to materialise for a variety of reasons. The classic here is probably Ashburton station, terminus of the branch from Totnes. The line was originally closed to passengers from 3 November 1958 but was to be secured for preservation by the Dart Valley Railway; unfortunately, the powers that be decreed that the trackbed between Bustfastleigh and Ashburton should be severed to allow for the construction of the A38 dual carriageway. The result was that, following reopening from Totnes to Buckfastleigh on 5 April 1969, the line reached Ashburton for only a brief period, the final train, seen here with ex-GWR 2-6-2T No 4555, running on 2 October 1971. Today, such is the importance of the tourist industry that, in similar circumstances, the road scheme would allow for the retention of the railway (as will be the case with the new road/rail bridge south of Grantown-on-Spey on the Strathspey Railway). Ironically, 30 years after closure of the link, the railway is now investigating the possibility of reopening from Buckfastleigh to Ashburton, but, one feels, it is a bit like closing the stable door ... *Ian Allan Library*

Below: Another Devon line to be the subject of a preservation bid was that from Barnstaple Junction to Ilfracombe, which closed to passenger services on 5 October 1970. After closure a number of schemes were proposed; indeed, the track was to be left intact for five years after pending the preservationists' raising the necessary funding. In the event, however, it all came to naught, and the line, like so many, passed into history. Here, on 23 August 1968, 'Warship' diesel-hydraulic No 814 *Dragon* awaits departure from Ilfracombe with the 15.20 service to Exeter. *David Birch*

It was not just steam that suffered from the syndrome of scrapping long after withdrawal. One of the long-withdrawn Class 42 'Warships', No 818 *Glory*, spent more than a decade as a fixture around the turntable at Swindon Works. Withdrawn in November 1972, the locomotive was to survive for exactly 13 years, until November 1985, when the pressure to clear the site resulted in its scrapping — another missed preservation opportunity. Fortunately (unlike in the cases of the NBR Atlantic and the HR 'Small Ben') two other examples of the type survive, but it seems unforgivable that no preservationist appears to have been offered the opportunity to save this locomotive. In happier days, on 5 November 1975, it is seen at its place of birth (and death). *Brian Morrison*

One of the most ambitious preservation schemes was that promoted for the much-mourned Waverley route. Under the name 'Border Union Railway', a campaign was launched to raise funds for the purchase of the complete route from Edinburgh to Carlisle. Although, by today's standards, the sums involved seem modest — a figure of £1 million being quoted for the purchase and reopening of the route — negotiations between BR and the BUR company failed, and the route was dismantled. In the decades after closure, piecemeal destruction of the route has taken place — the demolition of the viaduct at Hawick and the construction of a road on the trackbed at Melrose, for example — but, as described in the photo-section on the benefits of hindsight, the route is gradually gaining a second life. Here, on 20 April 1965, a BRCW Type 2 (Class 26) is pictured near Stave in Midlothian, amidst the splendid scenery for which the route was famous, with a service from Edinburgh to Carlisle. *J. C. Beckett*

If one event can be said to have sparked off the revolution in terms of public appreciation of the country's railway heritage, it was the wanton destruction, as part of the station's redevelopment in the early 1960s, of the dramatic Doric arch at Euston. Whilst few would have denied that the original station needed modernisation as part of the West Coast electrification scheme, the failure to incorporate the arch — and, arguably, Philip Hardwick's other impressive work (such as the Great Hall) — marked a watershed in appreciation of the nation's Victorian heritage. Its demolition was thus both a blunder — because it could have been saved — and a benefit, in that it served as a clarion call for the preservation of other stations; without the loss of Euston, pressure for the wholesale redevelopment of Liverpool Street and the lack of use for the offices at St Pancras might have resulted in the destruction of these buildings. The drama of the arch is well illustrated in this view *c*1900. *Ian Allan Library*

Of course, preservation also has its examples of schemes that progressed so far and then, perhaps as a result of over-ambition, failed. One of these was undoubtedly the West Yorkshire Transport Museum. Conceived as a scheme to create an integrated transport museum at Low Moor, allied to preservation of the Spen Valley line southwards, the West Yorkshire MCC-funded project started off with a hugely ambitious collections policy, including the acquisition of a Class 506 EMU from the Glossop line and the battery multiple-unit from the Railway Technical Centre at Derby. Unfortunately, the abolition of the County Council in 1986 plus the reluctance of the successor authorities to assume responsibility for the scheme saw much of the collection disposed of. The Class 506, regrettably, went for scrap, but the battery unit was saved and has now returned to its original home in Scotland. A streamlined project, known as Transperience, was the result, but this lasted barely a couple of years before it too failed. Here the Class 506 unit is seen stored in the museum's temporary store at Hammerton Street depot in Bradford on 2 November 1986; the battery unit can be seen in the background on the right. Whilst the project itself failed, without the WYMCC's funding it is unlikely that the latter unit would have survived, given that asbestos removal alone cost nearly £30,000. *Peter Waller*

Sectorisation

In 1986 British Rail management decided that there was too much flexibility in its locomotive arrangements, the feeling being that operations in this field needed to be put in a straitjacket. Thus it was that the locomotive fleet was divided into 'Sectors'. On 15 October 1987 at Ripple Lane yard the Sectorised freight 'company' Railfreight was launched. It was then decided that this was still too broad an organisation and needed to be made more narrow — probably the word used was 'focused'. It is a word which often arises in the minds of people who are myopic. In the run-up to privatisation, the Railfreight Sector was itself split into three 'shadow' companies — Mainline Freight, Loadhaul and Transrail. Each successive reorganisation cost money to set-up and

constituted another layer of management which had previously not existed. Each Sector had a different purpose and was allocated a certain number of locomotives. These locomotives could not be used by any other Sector — except by prior financial arrangement. Within each Sector, each train that had to be run was allocated its locomotive. If a locomotive failed it was very difficult to use any other locomotive because it had its own job to do. This development seems to have been an early experiment for the privatisation/fragmentation that was only a few years away. The newly Sectorised locomotives were given bright new liveries with fancy logos to distract attention from the nonsense of what was being done. British Rail was consistently criticised for

Above: One of the three pre-privatisation freight companies was LoadHaul. Here, on 22 May 1997, Class 56 No 56106 approaches Derby with a southbound train of steel empties. *Brian Morrison*

Left: During the 1980s the monolithic rail blue disappeared in favour of liveries for individual sectors and sub-sectors, and Railfreight, as the combined rail-owned freight business was known before its division, was no exception. Pictured in coal sub-sector livery at ICI Wilton, near Middlesbrough, on 14 April 1988 is Class 56 No 56122, which had just been named *Wilton — Coal Power*. *Dr L. A. Nixon*

Right: Another of the three companies was Mainline. Here Mainline-liveried Class 58 No 58001 is seen near Charing on 5 July 1999 with a train carrying spoil from Southern Water's flood-relief scheme at Hastings to the ARC quarry at Allington. *Brian Morrison*

The third of the trio was Transrail, which was arguably the most innovative in its attempts to generate new traffic. On 14 July 2000 Transrail-livered Class 37 No 37412 *Driver John Elliott* arrives at Crewe with the 10.07 First North Western service from Birmingham New Street to Holyhead; at this time FNW was forced to hire in locomotives and coaching stock for its services as a result of problems with its planned new rolling stock. *Brian Morrison*

being 'monolithic', while in fact its organisation allowed more flexibility in operating than that invented by the free-marketeers and their talk of 'choice'.

The madness of Sectorisation of the locomotive fleet was to a large extent cured on 24 February 1996 when Mainline Freight, LoadHaul and Transrail were sold to Wisconsin Central Railroad Co and amalgamated to form the English, Welsh & Scottish Railway (EWS). (The intention had been that the three freight companies should be sold off separately, but WC's offer for all three was obviously too good to refuse!) The paradox is that the locomotive fleet was fragmented by the nationalised railway, while, under privatisation (essentially a fragmentation of the railway), the fleet was once again consolidated under one owner; Wisconsin's Ed Burkhart had always disapproved of the fragmented nature of the new railway.

All that waste of money and time setting up a rigid, fragmented system which could have been better spent on improving railway infrastructure.

The Things They Said

An example of the loathing felt by the free-marketeers towards the public-service railway appears in Professor F. Blackaby's book *British Economic Policy 1960-1974* (Cambridge University Press, 1978), where he refers to British Rail contemptuously — and inaccurately — as 'part of the soft morass of subsidised incompetence'. In 1980 Alfred Sherman, Margaret Thatcher's speech-writer, described BR using the same lurid and untruthful imagery in a pamphlet which called for 'a Minister for Denationalisation'. The purpose of such a minister, according to Sherman, would be to 'dismantle the vast, parasitic [*sic*] apparatus of nationalised corporations whose actions can destroy all hope of recovery and the freedom and rights which depend on its early achievements.' At a press launch of said pamphlet, Sherman is quoted as saying: 'British Rail is sucking the nation's economic blood, and denation-alisation would see the end of most rail traffic in Britain. The demise of BR is a natural evolutionary development, because railways are an anachronism today.' In 1980, a year of depressed trade, BR's deficit was a mere £25 million — 1% of its gross income. And, of course, this was due in part to obligation to repay Government loans for modernisation. In *The Right Track — A Paper on Conservative Transport Policy*, then Transport Minister Norman Fowler wrote: 'Conservatives reject the idea that transport ought to be regarded primarily as a social service to which the taxpayer must be forced to contribute huge and continuing subsidies — the best way to ensure the public interest is to promote free competition and free choice.'

During the 1980s Margaret Thatcher's Conservative Government developed a policy of privatising public-sector industries but notably left the railways well alone; however, John Major had a romantic notion of a return to the days of the 'Big Four', and in the 1990s his administration duly turned its attention to BR. The White Paper on Privatisation consisted of 21 pages of large headlines, white space and a mission statement markedly thin on detail. What it was rich in was mindlessly fanatical market-speak: 'Our objective is to improve the quality of railway services by creating many new opportunities for private sector involvement. This will mean more competition, greater efficiency and a wider choice of services more closely tailored to what customers want.' The words, when analysed, actually mean nothing at all. They are merely the unfounded assertions of a political zealot with no understanding of the practicalities of railway operation.

A railway simply gives the opportunity to move from 'A' to 'B'. Trains leave the station at an advertised time and people get on them. What else can they do? Only one train can be standing alongside the platform at any one time, and only one company can operate that train. Where does competition come in, and how will more competition make the railways more efficient? Efficiency is a matter of skilled maintenance — and a full staff. Competition cuts costs — and staff — to make a profit and is therefore less efficient.

During the Parliamentary debates in October 1992 on the White Paper describing the privatisation plan, John McGregor, Minister of Transport, betrayed his sublime ignorance of railways by asserting that 'the railway is a business like any other and can be managed like any other'. Yet he went on to propose treating the railway like no other industry — by separating the head and limbs from the trunk. He referred to the plan as producing a 'coherent' transport policy when what he was doing was introducing incoherence. He talked of 'the disciplines of the market place', and one wonders what 'discipline' there was in a market place — a place where traders must, of necessity, buy cheap and sell dear. There is no discipline in a market — it is merely an hysterical rush to become rich or avoid becoming insolvent, as the chaotic scenes of the Stock Exchange, sometimes filmed on TV, show. British Railways, on the other hand, was always disciplined — by the very serious books of Rules and Regulations for railway operating and by time-honoured engineering excellence. But this was despised; on 12 January 1993 Mr McGregor told Parliament: 'The public know that the monolithic, old-fashioned, nationalised industries are not the best way of providing services efficiently and responsibly.' By contrast, Mr McGregor's coherent scheme of fragmentation would produce a 'world-class railway'.

A basic assumption of the privatisers, revealed in Hansard, was that, since people were complaining about the antiquated aspect of British Rail, all that was required to achieve perfection was to sell the railways to any group of investors. But it would be the same, denigrated railway. The White Paper extolled the virtue of setting up a 'market' in second-hand railway carriages. Richard Branson had told a Select Committee on Privatisation that he would repaint old BR coaches. The old carriages then became admirable because they had been repainted using 'private' money. The assumption was clearly one of 'State ownership bad — private ownership good'. Several Conservative MPs, whilst welcoming the demise of BR and the advent of the New Jerusalem, were also very anxious to know whether the State would still subsidise the private-enterprise train services to their respective constituencies.

Mr Matthew Banks, the Conservative Member for Southport, said that 'if the objective is to create better quality services through the gradual introduction — I stress gradual — of private sector competition then [the Bill] has my 100% support'. He had not stopped to wonder how two companies could each run the 8am commuter service to Preston. He strongly implied that running a railway was simply a matter of 'providing more opportunities and choice'. Several Members repeated this phrase, absolutely irrelevant to successful railway operation, as if it were the entire content of their faith. Mr Banks was, however, way behind the plot, because he went on to pour scorn on those members of the House who seemed to think that his Government was 'trying to privatise British Rail lock, stock and barrel'. Such a course, he informed the Deputy Speaker, 'would be a ridiculous and impractical course of action'. But that is what happened.

The British taxpayer paid £629 million just to get the Privatisation Act on the Statute Book. To encourage the risk-taking free-marketeers to buy Railtrack, £1 billion of BR debt to the taxpayer — which Railtrack would have taken up in any normal sale — was wiped away. In BR's Annual Report for 1993/4 the land, buildings and infrastructure were valued at £6.4 billion. Railtrack paid the taxpayer £1.9 billion for these assets; this was just enough to cover the taxpayer's subsidy to the Train Operating Companies in 1994/5 of £1.74 billion. In 1993/4 BR's subsidy from the taxpayer had been £930 million.

The Results of Privatisation

It used to be a cliché, but it was a truism — painting the Forth Bridge was a job that would (and should) never end. However, this was not a policy to be wholeheartedly endorsed by Railtrack when it announced a temporary moratorium on bridge repainting. This proved to be another of Railtrack's PR blunders, as the condition of the bridge seemed to deteriorate rapidly, and the painting gangs were soon back on duty. In happier days, on 13 September 1985, a three-car DMU heads northbound towards the Fife shore. Evidence of painting can be seen in the scaffolding and patchwork of different colours.
Ian Allan Library

The scheme devised and put into effect for the privatisation of railways was 'sold' to the public — although a large section of the public never 'bought' it — as a way of saving money and improving the performance of the British railway system. The entire exercise was a blunder, and the root cause of this was the idea that railways were a business rather than an investment in raising the quality of life in the country as a whole. The result has been that railways cost the public more now than they did under state ownership.

The Minister of Transport assured Parliament that the new arrangements were 'flexible and practical'. They were certainly flexible, and no sooner had they been implemented than they took up a shape similar to that of a pear. The Minister was sure that the private sector had the 'flair and enterprise' to run the railways better than the (by implication) dull and unenterprising men and women who had been paid by the Government to run them. Later in the same statement the Minister said that he would 'encourage management buy-outs'. But these were the same people he had just criticised.

The Government divided routes into 'franchises' and invited a competition from interested parties — which consortium would (to borrow a phrase from Brunel, when invited to enter a competition to become Engineer of the GWR) 'make the most flattering promises' to the Government in return for a licence to run the trains. Legally binding 'safeguards' of the public interest abounded, of course. The franchisees would take on an annually decreasing subsidy until they were paying the Government out of the profits they would naturally be making because of their 'private enterprise'.

Anglia Railways is a franchise formed in 1996 to run the services from Norwich to London, Ely, Cromer, Yarmouth

Above: As a result of the declining freight business in the late 1980s and early 1990s, the number of new locomotives put into service by BR declined dramatically. For almost 150 years, railway-owned workshops had been constructing locomotives; the last diesel locomotive to be built by British Rail in its own workshops was Class 58 No 58050, which emerged from Doncaster Works in March 1987. BR's next new diesels, the Class 60s, would be constructed by Brush at Loughborough between 1989 and 1992. Here No 58050, now withdrawn and destined for preservation, is seen at Toton on 5 September 1987 when relatively new, shunting Class 45 locomotives destined for the scrapman. *N. E. Preedy*

Below: The tradition of BR Works-built locomotives came to an end with the construction of No 91031 *Sir Henry Royce* at Crewe in February 1991. Subsequent electric locomotives, of Class 92, would also be built at Brush. For the domestic locomotive-building industry, the investment hiatus caused by privatisation was serious, and, when the decision was made by Railfreight's successor, the English, Welsh & Scottish Railway, to invest in 250 new disel locomotives (Class 66), the work went to General Motors in Canada. Since privatisation Britain's main-line operators have taken delivery of more than 300 new diesels, none of which have been built in this country — a sad reflection of the decline of the nation's manufacturing capacity. Here No 91031 enters Durham on 28 June 1991 with the 11.30 from King's Cross to Glasgow, which would actually terminate at Edinburgh. *Ian S. Carr*

and Lowestoft, and from Lowestoft to Ipswich. One of the directors of GB Railways, parent company of Anglia Railways, is one Michael Schabas. Early in 2000 he gave a speech to the Confederation of British Industry and showed his grasp of railway matters by criticising a journalist who had said that punctuality was the most important thing on a railway. The wonderful ways of the Strategic Rail Authority are such that they allow Train Operating Companies (TOCs) to omit the worst punctuality days from their punctuality tables. BR was never allowed to cook the books in that (or any other) way. During the last three months of 1999 Anglia Railways was able to produce punctuality figures from which no fewer than 11 days of bad news had been omitted — and, even then, its punctuality fell short of the target set in its so-called Passengers' Charter. Mr Schabas continued to show his grasp of the railway by telling the CBI members that: 'Responsibility is now fragmented, which means that instead of one big boss making decisions one at a time, slowly and usually wrong, there are thousands of people making commercial decisions on a daily basis. Nationalised monopolies just count up the losses; this appeals to a certain kind of person and those people resist change very hard.'

In 2003 Mr Schabas's express trains are formed mainly with British Rail coaches, hauled largely by 38-year-old, BR-designed electric locomotives. They draw their power from overhead wires erected at the expense of British Rail (and thus, ultimately, the British taxpayer) and are signalled by modern signalling, also paid for by BR — surely not a wrong decision by BR in 1980? Using these assets, he counts up his profits which have only become profits because of the subsidies his franchise receives from the public purse.

The only changes Anglia Railways can make to its service are cosmetic or else to offer cheap fares which amount to running at a loss. The reason Anglia is powerless to control its destiny is that it does not own the carriages, or the track, or the signals that keep them safe. Unlike BR, which controlled all aspects of the railway, Anglia Railways cannot alter the timetable, put on additional trains, make better scheduling arrangements or go outside their franchised area without lengthy consultation with other agencies. This unprofessional way of arranging railway matters leads to a great deal of inconvenience for passengers. East Anglia, once under a single authority over whose tracks trains could run as far as they liked, is now fenced off for Anglia Railways, West Anglia Great Northern, First Great Eastern and Central Trains — whose interests will sometimes conflict. After Anglia Trains took over its little fiefdom, trains from Norwich could no longer pass Ely — a situation unknown since the line was opened in 1845. Passengers for Cambridge had to get out at Ely and await the entirely unconnected trains of the WAGN franchise.

On 25 January 2002, with a great deal of media hype,

In its near 10 years of existence, Railtrack received enormous criticism as a result of the perceived state of running lines. The failure to ensure the adequate clearance of weeds was one of the most directly evident signs of a failure, in the eyes both of enthusiasts and of regular passengers, to maintain adequately the infrastructure inherited from British Rail. Whilst the running of weed-killing trains under Railtrack did not cease, there was an undoubted — and perhaps justified — perception that the weed growth was symptomatic of a deeper malaise within Railtrack's stewardship of the nation's railway assets. *Brian Morrison*

Another facet of the maintenance problem was the spread of graffiti afflicting not only rail-side structures but also the facilities at passenger stations. Whilst the railway industry cannot itself be blamed for a wider social ill, the failure to clean graffiti and to take action whenever possible against perpetrators may well have increased the perception amongst more vulnerable groups that railway stations and railway travel in general was less safe than previously. This view shows the impact of graffiti upon Mottingham station in southeast London. *Brian Morrison*

Anglia Railways announced that, from October 2002, it would be running an hourly service from Cambridge through to Norwich. This was trumpeted on local television as a major triumph. Any change that brings the modern service back to what it was in BR days is a triumph, but it is no triumph for private enterprise — it has been done on public money. To run the new service Anglia Railways will lease four brand-new trains, and to pay for them the Strategic Rail Authority will pay Anglia Railways £10,000,000 from public funds for its 'private enterprise'. In May 2002 the Anglia Railways franchise had to be propped up by an additional subsidy from public funds of £23.7 million, of which £3.2 million was paid at once. Also in 2002 the Strategic Rail Authority (SRA) agreed to increase the subsidy to be paid to Central Trains over the period 2001-4 — when the subsidy should have been reducing — from £445 million to £500 million, while ScotRail had its subsidy for the same period increased from £702 million to £772 million; should either of these franchises make profits above a certain level, it will pay the excess to the SRA. All this seems to suggest that privately owned railways are less capable of making a profit than was InterCity in the 'bad old days' of British Rail.

When the railway was one great organisation, workers joining it expected to spend a lifetime in its service and felt they were part of a team. Under privatisation everything fragmented, and old hands became known as 'dinosaurs' — 'management-speak' for loyal, experienced railwaymen. When track-maintenance work is carried out by contractors, they feel no commitment to the railway, because they are not employed by the railway and probably have little experience of the teamwork required. On Sunday 3 November 2002 at Aldwarke Junction, on the line between Masborough and Doncaster, the employees of contractor Jarvis Rail were working on a set of points

(4256). The 'V' in the crossover from down main to up main was faulty and had to be removed. The men unbolted the 'V' rail and lifted it out and took it away. They then replaced the 'V' rail with a length of straight rail so that trains travelling straight along the down main would have uninterrupted rails to run on, but there was a gap in the rail forming the switchblade to take the turn-out at the crossover. It was the responsibility of the men doing this to ask the signalman before they started the work so that electric power to the point motor could be cut off. Sheer common sense should have told them that, having created a gap in the rails, they should advise the signalman before going away. There was then a large gap in the crossing rails. They ought to have put a point clamp on the switchblade to prevent it moving, but they did not. They were not railwaymen, just men hired by the contractors. They apparently had no understanding of what they had done. The points stayed in that condition for five days. It seems the patrol man did not question the situation, even though there was no clamp on the points. Perhaps there was no patrol man. At 1730 on 10 November the signalman powered the points for a crossing movement, and at 1735 Freightliner locomotive No 66507 and two wagons were derailed where the rails ended. The locomotive and wagons were re-railed at 02.30 on the 11th, and the line was reopened to traffic at 18.17. Trains operated by Midland Mainline and Arriva Trains Northern were diverted, those of Virgin CrossCountry were suspended, while buses ran between York and Derby; passengers travelling from Cleethorpes to Manchester went by bus from Sheffield to Rotherham. Just another day in the history of the privatised railway, where this sort of nonsense happens so regularly that real railwaymen's morale has to be numbed in order to cope with the shame, frustration and embarrassment.

Prior to Privatisation, BR had established a niche market for express parcels traffic under the trading name of 'Red Star'. As with other subsidiaries, this was sold off to the private sector and would ultimately disappear — another loss of traffic to the railway industry as a direct consequence of the fragmentation of the industry. No doubt the traffic is still there; today, however, it is more likely to be carried by couriers, adding to the congestion of the country's already overcrowded road network.
Ian Allan Library

Above: For many years the important junction on the West Coast main line at Rugby provided a sorry sight to passengers, with its LNWR-built overall roof in a sad state of repair and its red-brick structure showing clear evidence of under-investment. A major refurbishment scheme saw the original trainshed demolished and new platform canopies erected alongside a repaired main building, but no sooner had the work been completed, in mid-2003, than it was announced that as part of the West Coast main line modernisation scheme — itself a classic blunder, in terms of cost and time over-run — the newly-refurbished station was to be swept away and replaced by a new layout designed to provide faster through running lines. At a time when the railway industry is increasingly coming under the spotlight in terms of investment funding, this lack of joined-up thinking seems feckless in the extreme. Here, on 1 July 2003, the refurbished station plays host to Class 66 No 66508, Class 86/2 No 86210 *CIT 75th Anniversary*, both stabled, and Class 321/4 EMU No 321413, forming Silverlink's 08.48 Euston–Birmingham service. *Brian Morrison*

Below: As part of the privatisation process, 25 Train Operating Units (TOUs) were established and put out to franchise as Train Operating Companies (TOCs). The franchises varied in length and in financial terms, but each of the TOCs was set targets in terms of performance, both financial and operational, with the ultimate sanction that failure to meet these targets would result in the franchisee losing its franchise. So far, two franchisees — both controlled by Connex — have been stripped of their respective franchises, Connex South Eastern losing its rights in 2003. On 7 April 1999 the 12.05 London Victoria–Ramsgate service, formed of Class 365 units Nos 365505/13 in the Connex livery of blue, yellow and white, passes Shortlands. *Brian Morrison*

Automatic Train Protection

The train-control system known as Automatic Train Protection (ATP) is a fine example of why railways should be self-contained, with all engineering services provided internally by people who remain in railway service for many years and gain experience. The railway should have a stable, long-service, public-service ethos and be directed from the top down by a single, central control. Railways did their finest work for the nation when thus organised. The greatest blunder ever made on railways was to turn that traditional concept inside out, back to front and stand it on its head. But in today's 'retail' world, 'traditional' is a dirty word — unless it is used as hype for a tourist attraction, in which case it is OK.

In 1985 Major C. F. Rose, Chief Inspecting Officer for Railways, reported the increasing number of signals being passed at Danger and said that British Rail must install a more powerful system of train control than that provided by the existing Automatic Warning System (AWS), which had its origins in the mid-1930s. Although the terrible tragedy at Clapham on 12 December 1988 could not have been prevented by ATP, Sir Anthony Hidden's report on the crash recommended the fitting of an ATP system 'within five years'.

In March 1989 there were two ATP-preventable collisions, costing a total of seven lives, at Purley and Bellgrove on the 4th and 6th respectively. On each occasion, the BRB declared its full commitment to the huge project of developing and fitting ATP. But the BRB did not control expenditure — the Government did.

By the late 1980s SNCF (French National Railways) had installed on its new, high-speed lines a powerful system of train control known as TVM, whereby the speed of the train is continuously monitored by the interaction of track- and train-mounted equipment. Should speed at any point be excessive in relation to the signal aspects showing or to the maximum line speed, the train's brakes will automatically be applied after the driver has been given an audible and visible warning. This is a continuous-monitoring Automatic Train Protection system. The system adopted for Britain is intermittent. Information is sent from induction loops or beacons lying at intervals in the track and is picked up by the antennae on the train. It cannot, in fact, prevent a driver from passing a signal at Danger, and it also allows drivers to approach buffer-stops at 10mph. After passing a single yellow the train speed is controlled at anything from 10mph to a maximum of 50mph. There is nothing to stop the train passing the next signal at Danger at the controlled speed, but as it passes over the beacon the brakes are automatically applied to bring it to a stand — theoretically — before it can do any damage. There can also arise the ludicrous situation whereby the driver can see that the next signal is green but, because he has received a warning from a single yellow, is unable to accelerate, because the ATP equipment 'thinks' the next signal must be at red — until the next loop or beacon informs it otherwise.

The system chosen for trial in Britain was complex in comparison to conventional BR AWS and requires more advanced design and closer tolerances in installation when setting up the beacons and loops on the track. When the system was first put on trial, permanent-way maintenance staff were railwaymen and better able to cope with the demands of higher technology, but, with privatisation, contractors were employed whose staff was increasingly inexperienced. It took a long time to educate the newcomers that the induction loops laid down the centre of the sleepers were there for a purpose and not to be damaged by rough handling. This remains a major problem with staff employed by private contractors, who might break the wires through ignorance of their importance or leave piles of ballast in the 'four-foot' which prevents the reinstatement of induction loops after track repairs. Network Rail procedure on ATP induction loops even allows for the loops not to be reinstated immediately after re-laying, because of the difficulties in getting staff sufficiently skilled to do the job; as a result, the ATP system lies with gaps in the protection it affords, and train delays are accepted as inevitable.

Attempting to fit the very 'hi-tech' ATP to an increasingly run-down and understaffed British railway was never going to be easy, but, as ever, the Government and the media would demand perfect safety from a railway towards which they had always exhibited the greatest

When the French developed their high-speed network for the operation of the fleet of TGVs at 300km/h (186mph) it became evident that conventional signalling equipment would not be appropriate for such high-speed running. As a result the TVM system was developed which replaced conventional signalling with a computer-based system. This has proved to be successful. Here SNCF TGV units Nos 75 and 68 are seen near Cluny on the St Etienne section of an express from Paris on 30 September 1984. *John C. Baker*

parsimony and even outright animosity. Meanwhile, no induction loops were fitted in motorways to slow down speeding road vehicles.

In January 1990 the BRB set up an ATP Project Management Team and awarded contracts to two companies, the Belgian ACEC, with its TBL system, and Standard Electric Lorenz, with its SELCAB system. There would be a two-year trial, after which one of them would be selected for a further, 10-year trial. In 1991 the SELCAB system was fitted to 210 signals on the Marylebone–Bicester North/Aylesbury suburban and outer-suburban route; the TBL system was installed on 455 signals on the high-speed Paddington–Swindon–Bath–Bristol line. The 1950s AWS system meanwhile remained the standard warning/braking device.

Just as the BRB had established an organisation to see the job through, and with the contracts for the two pilot schemes signed, a major alteration to the structure of the railway took place. In April 1990 the BRB, uncannily anticipating what was to come, abolished the traditional Regional and Departmental structure of the railway — the last link with a saner, more engineering-orientated railway world — and adopted an Orwellian example of double-speak: 'Organising for Quality', with the official but unfortunate acronym of OfQ. The railway was now split into five 'Sectors': InterCity, Provincial (later changed to Regional Railways), London & South East (later changed to Network SouthEast), Railfreight and Parcels. These 'businesses' were then further sub-divided into 'Divisions'. More Orwellian terms appeared: 'Corporate Leadership 500' and 'Leadership 5000'. The titles of the new 'businesses' were aptly written back to front — 'InterCity Great Western' or 'Network SouthEast'. Passengers became 'customers', and long-service railway-men, Stationmasters, experienced in operating, became 'Retail Managers' — if they stayed at all. The detail of reorganising (or disorganising) under OfQ took two years and was completed in April 1992 — the precise period during which ATP should have been developed. The complex project required a coherent, all-in-one management structure, but OfQ was the opposite of that. The reorganisation cost a mere £150,000,000, which, had it been spent on the track instead of deckchair-shifting, could really have raised the level of quality on the railways.

The ATP project came under the control of InterCity Great Western (ICGW), Brian Scott, formerly Western Region General Manager and now Chairman of ICGW, being appointed Chairman of the Steering Committee trying to progress the installation of ATP. His task was made much harder because he was in charge only of a train-operating organisation rather than a railway and had to gain the co-operation of independent organisations concerned only with what was profitable for them. The curious fact that the Health & Safety Executive (HSE) did not believe that ATP was worthwhile meant that other train operators felt no urgency to spend money on it. ICGW would make costly efforts to install the equipment while Network SouthEast ran over the same tracks with non-ATP-fitted trains.

By March 1993 472 ICGW drivers were fully trained and 46 High Speed Trains (HST) power cars had received the ATP, but only a very few signals had been fitted with the necessary beacons or induction loops. On the Chiltern line the tracks were 95% fitted with ATP, but there were no ATP-fitted trains and few trained drivers.

A large number of ICGW drivers volunteered for training with ATP, but then the lack of a significant amount of ATP in service was disappointing to men who had had their expectations raised by their training, and morale fell accordingly. Where ATP was installed, it did not always work properly, and there were occasional acts of vandalism perpetrated on the cab equipment by disgruntled drivers. When ICGW started to cut its costs by offering voluntary redundancy to drivers, 20% of those who had trained — senior men — gladly took the money and ran. Yet other trained ICGW men accepted the lure of higher wages from other Train Operating Units and left.

In 1993 the Government, through the Health & Safety Executive, asked for a cost-benefit analysis of ATP to be undertaken. The figures were presented to the HSE, and the HSE, dedicated to a safer environment, recommended that the benefits of ATP were insufficient to justify the cost of nationwide installation. (The costs of gigantic earthworks on motorways are, of course, calculated using a different set of values.) The cost, in 1994, of ATP was authoritatively estimated at £475 million. In 1995/6 the subsidy which the Government paid to the new owners of the railways was approximately £1,160,000,000 more than had been paid to the BRB in its last year. Clearly, the value of human life was a drop in the ocean compared to what the Government was prepared to pay to sell the railways. Another way of comparing the

supposed vast cost of ATP is to compare its £475 million with the £600 million (on conservative estimates) of public money that the Government spent to sell the railways. Or one could look at the £5 million that it would have cost Thames Trains to install ATP on its trains to avoid the accident at Ladbroke Grove and the £7.5 million the same company paid to its shareholders in the two years after the crash. The HSE having turned down a reasonably priced system which would, more or less, have guaranteed safety on the railways, British Rail had little choice but to abandon its commitment to ATP — although the two pilot schemes were to be completed.

The railwaymen of the ATP project team persevered in spite of the 1993 HSE finding, and in 1994 they had 87 HST cabs fully fitted with the equipment, as well as 34 Class 165 DMUs on the Chiltern line. Owing to the delays caused by 'Organising for Quality', the ATP system was not even then fully operational, because there were unresolved antennae and vibration problems with the train-borne equipment, and still not all the trackside equipment had been installed.

Besides the new obstacles there was the problem of safety certification. ATP had never been formally accepted by BR. And Railtrack, as proprietor of the railway tracks and signalling, had to be able to certify its safety. Although the ATP would definitely prevent drivers from passing signals at Danger and although it was basically the same system as was then being used on other European railways, it was still considered necessary that it have a Railtrack Safety Approval Certificate. The difficulty here was that in 1994 Railtrack did not possess sufficiently experienced staff to carry out a risk-management assessment on the ATP equipment. Indeed, 'safety guidelines' had yet to be formulated before any risk assessment could be submitted to the test. When these guidelines were eventually thrashed out and the ATP safety case had been tested against them and approved by Railtrack, the approval of the HSE had to be obtained. All that Railtrack then needed was the money to carry on with the work, which (in 1995, at any rate) was not forthcoming.

The most delicate parts of the ATP on-train equipment were the tachometer and the under-train antennae, the latter communicating with the beacons and induction loops in the track. The antennae had to be fixed at a precise point under the power car, and squeezing them in provided a great headache for design teams. Having consulted the original drawings of the HST, designers prepared a modification design and set of drawings to show how and where the ATP equipment would be fitted and how and where the circuitry should be installed to integrate ATP with the train's braking and electrical systems. When the drawings went out to the depots to be implemented it was discovered that the HST power cars were no longer standard in their detail — they had been extensively modified at various depots around the

The Class 165 DMUs were designed to replace ageing first-generation DMU stock on the lines to the north of Marylebone, and the trials and tribulations of their introduction into service (with ATP) are well recounted in the narrative. Fortunately, however, they have proved to be successful in service and have helped the Chiltern Trains TOC achieve considerable success in terms of the Passenger Charter and in the reinvigoration of the once-threatened line into Marylebone. Here, on 11 April 2003, No 165034 stands at Aylesbury prior to forming the 17.49 service for Marylebone. In the distance can be seen refurbished Class 121 single car No 121120, which will form the 17.55 to Princes Risborough. *Brian Morrison*

country. Often the modifications had been shoddily executed, and certainly 'Head Office' had no knowledge of the alterations that had been made. The result of this was that before the HSTs could be fitted with the ATP gear they had to be re-wired to the standard design upon which the new drawings for the ATP installation had been based. This was another excellent example of the need for strict, central control on railways.

The necessary HSTs having been re-wired and variously brought back to original specification, the ATP gear was fitted and trials began. It soon became obvious that the site for the on-train equipment had been badly chosen, as had the method of fixing: the antennae and tachometer were shaken to bits and were unable to perform their function.

Class 165 diesel multiple-units destined for the Chiltern line had been designed and were in the earliest stages of construction when the decision to fit them with the ATP equipment was taken. This led to a delay in the construction while the same modification problems and the matter of builder's extra costs were resolved. This also delayed driver training, but when a few APT-fitted '165s' became available training began, first for Chiltern-line managers and then for the rank-and-file drivers. By March 2003 100 men were APT-trained. Unfortunately there were still insufficient Class 165s for them all to practise on — with or without ATP.

By the time the trains had been delivered, the Chiltern Train Operating Unit was most anxious to get them into revenue-earning service, so the ATP project team had to compete — and there's nothing so good as competition on railways — with the train-operating department for access to the units for the purpose of installing the ATP on the '165s' at the depot. It was then discovered that the quality of the new vehicles was such that the units had to be extensively modified at the depot to ensure that the passenger service obligation could be met, and there simply was not time to 'add on' the ATP — even though the Chiltern line was one of the two National ATP Pilot Schemes to which the BRB was committed. When the ATP team was eventually able to make a start on installing the on-train ATP equipment they discovered that the builders had not installed vital fixtures, and the relations between the manufacturer and the other parties involved became so soured that the manufacturer threatened to get out of the scheme altogether. Eventually, by June 1992, a few ATP-fitted '165s' were ready, but service conditions showed up tachometer-bearing failures, drive-assembly failures, inadequate software for the on-board computers, shortage of trained drivers, and a fault in the braking performance. The brakes on the '165s' were found to be not according to specification — the specification on which was based the design of the ATP — being too powerful and requiring modification before the ATP could be expected to work properly.

There was a shortage of skilled men to install and test ATP because during the run-up to Privatisation — which was to deliver a 'world class' railway — hundreds of technical staff had made redundant — a truly Orwellian concept of Quality — and also because major signalling work on the London–Folkestone–Channel Tunnel line had corralled a large proportion of remaining BR-trained technical men.

The ATP induction loops lying on the sleepers were vulnerable to lightning strikes (*ie* heavenly lightning, not 'industrial action'). ATP installation had then to be suspended while these intricate electrical problems were attended to. To obstruct progress still further, the work, having been installed, could not be 'signed off' except by high-grade testers, who were also in short supply, for reasons given above.

Then came the most cataclysmic disaster the railways have ever faced, with effects worse than the intervention of the Luftwaffe in the years 1940-2 — the Railways Act 1993. This permitted the total fragmentation of the system and then fragmentation of the fragmentation. It was 'every man for himself' on the railway. The ATP project, of necessity, was lost from sight as members of the project team concentrated on their own job security and the 'Corporate Leadership 500' busied themselves with organising lucrative 'management buy-outs'. Railtrack took over the BRB's formal commitment to ATP and responsibility for the installation of the ATP on the track, but that part of the equipment which might eventually be housed on board the trains was to be managed by the Train Operating Company (TOC), although the TOCs were not required to commit themselves to installation. The train on which the equipment was installed belonged to the leasing company which owned the vehicles. During the process of privatisation there was no budget to fund the ATP work. Instead of being a railway, organised to provide trains for the public, it became a kit of parts to be bought and sold. The companies which had provided the equipment for the two pilot schemes were sold and sold again: ACEC was merged with GEC-Alsthom, which then changed its name to Alstom; Standard Electric Lorenz sold itself to Alcatel,

which was then bought by Alstom. Even the BRB Project Group was sold — to Transportation Consultants International (TCI) and then again to Atomic Energy Authority Technology, becoming known as 'AEA Technology — Rail'. All this conspired to delay the installation of ATP, and with the delays came increased costs, making even the pilot schemes less attractive to the eager new businessmen who had once been railwaymen.

By the end of BRB control, in 1994, ATP had been installed on the Chiltern line between Marylebone and Aylesbury via High Wycombe, between Amersham (LT excluded) and Aylesbury and between Princes Risborough and Aynho Junction; on the Great Western main line it had been installed between Paddington, Bath and Bristol (North Somerset Junction), with some deliberate gaps to give drivers experience of passing into and out of ATP control, and between Wootton Bassett and Bristol Parkway. The 'deliberate gaps' will be filled in the foreseeable future — the money is now available.

ATP has had the greatest degree of success on the Heathrow Express line, where the train-borne ATP equipment was designed into the brand-new trains. The tracks from Paddington to Airport Junction were renewed, and, although this was not done specifically for the fitting of ATP equipment, it helped to have new track for installing the new technology. The task on other, older trains was made very difficult by the necessity of fitting electronically sensitive equipment to vehicles which were not tailor-made to receive it. The equipment had to be manufactured in the order in which it was required on the trains and in the track. Trains had to be fitted with one half of the system, signals and lineside equipment with the other. Skilled technicians had to be trained to do the installation and drivers trained in its use. A vital component, a tachometer, must be fitted to one wheelset. On the Western Region, wheels on the 1976-built HSTs sometimes slipped during acceleration or skidded during braking, in either case entirely confusing the tachometer/computer link. Vibration during running also interfered with the working of the new equipment, grafted onto the old trainsets.

Because the railway is now operated by individual companies which please themselves what equipment their trains shall carry, the problem arises of what automatic braking system to adopt — AWS, ATP or TPWS (Train Protection & Warning System). The objective of TPWS is to take control of a train and apply the brakes when approaching at excessive speed a signal at Danger. It does not provide total protection from passing signals at Danger — but neither does ATP.

In 2002 the decision was taken by the SRA to fit 11,000 signals with the TPWS system. Even those lines already equipped with ATP will get TPWS; the Chiltern line is ATP-fitted, as are 99% of the trains running over it, but TPWS is being installed in addition to ATP. The HSTs leased by First Great Western are fitted with AWS, TPWS and ATP.

The Three-phase EMU

It was in 1991 that the prospect of Trans Manche Super Trains (TMSTs) running on BR came to light; for use on international Eurostar services, these would consist of 20 coaches, with a total of 12 powered axles. Contemporaneously, new types of domestic EMU were being designed which would use three-phase electricity. It was realised then that the power supply was inadequate for these new trains, and monolithic, rigid, poorly cultured BR began to plan the modernisation of the electricity-supply equipment to accommodate the increased demand. Then came Privatisation — the brain-child of those who believe that the Market — supply and

demand — will solve all problems. The result was that the engineers who had been working on the power-supply problem were either made redundant or dispersed into the employment of the new companies, and the modernisation of the power supply was still-born. But this did not prevent the newly formed Train Operating Companies from ordering lots of new EMUs.

The 'Eurostars' were introduced in 1993, while the Class 365 'Networkers', built at York, appeared in 1994/5. They are all powered by three-phase electric traction motors. Much more efficient than old-fashioned DC, as used with great success for over 100 years by the old private companies and the much-denigrated BR, three-phase electricity can be computer-controlled to make the motor deliver a range of power far beyond the capability of the old English Electric motors of yore. This is excellent,

Eurostar sets 3303/4 departing King's Cross with a special service to York on 24 July 1999. This service was the precursor to the use of these units on services to Leeds. Brian Morrison

Above: Class 365 No 365507 stands alongside Class 375 'Electrostar' No 375617 at London Victoria at 10.29 on 30 April 2001. The former had formed the rear of the 08.15 service from Ramsgate and Dover, the latter the rear of the 08.15 'shadow' service from Ramsgate. This was the first day of 'Electrostar' services to London.
Brian Morrison

Right: Pictured between Rainham and Newington, two Class 375/6 'Electrostars', Nos 375617 and 375625, form the 13.25 service from Gillingham to Dover and Ramsgate on 10 April 2001.
Brian Morrison

Above: The Class 334 'Junipers' were designed for operation with ScotRail, replacing older EMUs on services funded by Strathclyde PTE. However, the class's entry into service has been much delayed. Here unit No 334029 is pictured at Newton-on-Ayr with the 14.00 service from Glasgow Central to Ayr. *Brian Morrison*

Right: Amongst the most stylish of the new generation of EMUs are the Class 460 'Junipers' designed for use on the Gatwick Express service from Victoria to Gatwick Airport. Here No 460003 approaches Gatwick with the 14.15 service from Victoria on 25 June 2001. *Brian Morrison*

Also known as 'Junipers' are South West Trains' Class 458 units, one of which, No 8025, is seen leaving Waterloo with the 13.12 service to Basingstoke on 14 March 2002. Early units of this type suffered from harsh riding characteristics, faults with the braking system and excessively noisy air-conditioning apparatus. SWT should have been better served for its £100 million investment and duly changed its supplier, but this does not address the problem of power supply anticipated when all its trains are worked by three-phase EMUs.
Brian Morrison

but it might have been a good idea to ensure first that the power supply would be adequate.

Neither the LBSCR nor the LSWR, neither the Southern Railway nor British Railways, introduced trains for which there was no power supply. These new trains are highly allegorical of the new capitalism: they suck up so much current that they rob passing trains of their power and overheat the electricity sub-stations. In what way can 'market forces', 'choice', the 'disciplines of the market place' assist in this matter? Will those who used this language to denigrate British Rail now come forward with the necessary cash to install a modern power supply commensurate to modern traction requirements? I don't think so. Who would do the job? To supply the heavy flow of cash would drain resources from elsewhere and melt down the value of their shares.

Euro Sleepers

John Prescott was applauded by the Press for saving the Channel Tunnel Rail Link by supporting a unique financial deal. By telling investors that the Government is backing the project, London & Continental Railways is getting the money at a lower rate of interest. The private sector will provide £3.7 billion and the Government a subsidy of £130 million.

The Link will be built in two stages. The first will be Channel Tunnel–Fawkham Junction, near Ebbsfleet, ready in 2003, but the bad news is that the Government has been forced to write off the cost of the Eurostar Nightstar international sleeper trains at a cost of £10 million and accept that they will never run on the L&CR tracks.

British Rail ordered 139 coaches — 72 sleeping cars, 47 saloons and 20 lounge/service carriages — intended for Eurostar (UK) services from Plymouth, Swansea and Glasgow to Paris and elsewhere in Europe. The carriages were built by GEC-Alsthom at a cost of £1 million each. By the time 20 had been fully equipped and 80 more partially completed, British Rail had been sold off. The new company to run the international services, European Passenger Services, did not want the carriages, so the Government was obliged to pay GEC-Alsthom for its work. Having paid £10 million, the Government took no further interest. The carriages were taken to the MoD sidings at Kineton, beside the old SMJ line, in rural Warwickshire, and left in the open to the tender mercies of wind and rain. John Prescott himself said: 'It is a bloody scandal. I have had to pay £10 million for sleeper trains sitting in a field. I could have done a lot of much more useful things with the money.'

Destined never to turn a wheel in passenger service in Europe, unused 'Nightstar' coaching stock is pictured stored in the Works Yard at the Alstom complex at Washwood Heath on 17 November 1998. All the surplus stock would eventually be sold — at a hugely discounted price — to the Canadians. *Brian Morrison*

Above: 'Nightstar' was not the only investment in European services that ultimately proved to be a partial waste of money. Delivered along with the full rakes destined for London–Paris/Brussels services were a number of shortened rakes for services linking Europe with places like Manchester, Leeds and Edinburgh, running over the West and East Coast main lines. In the event, a change of policy meant that the concept of North of London services was abandoned before the units entered service. A new home for a couple of rakes was, however, to be found with the main ECML franchisee — Great North Eastern — which used them to supplement its existing stock on services between London and Leeds. The one downside, however, of this creative reuse, is the amount of 'dead' mileage required to link King's Cross with the units' maintenance facility at North Pole depot. Here unit No 3305 awaits departure from Leeds on 7 January 2003 with a service for London. *Brian Morrison*

Below: In order to service the North of London units, huge investment was made in servicing facilities, including this depot near Manchester Piccadilly. Unfortunately, despite the slogan, it would be a case of *'le Eurostar n'habite pas ici'*! Never required for European services, after some years with no definable role the new depot will be utilised for the maintenance of the new 'Pendolino' units being introduced on the West Coast main line. *Chris Dixon*

Above: As a trial for North of London services, British Rail introduced a 'shadow' service from the North using HSTs on a through service to London Waterloo. Not only did this service prove the lack of a market for through trains from the North; ticketing restrictions, which meant that one had to be in possession of a Eurostar ticket covering cross-Channel travel before one was able to use this service, meant that the HSTs regularly operated with barely a handful of passengers. Here an HST set, with Nos 43146 (leading) and 43129(rear), departs from Waterloo with the 12.45 service to Edinburgh on 5 February 1996. *Brian Morrison*

Below: A full house of Eurostar services at Waterloo International on 4 June 1999. With its stunning (if problematic) roof, this was the first London terminus to provide uninterrupted international services. However, with the construction of the second phase of the Channel Tunnel Rail Link, the vast majority of Eurostar services will be transferred to St Pancras, making Waterloo International less important and also reducing the attractiveness of Eurostar services to those who live south and west of London. *Brian Morrison*

Cranes No More

At 8.19am on 8 October 1952 at Harrow & Wealdstone station a triple-train crash occurred. An up express from Perth crashed into the rear of a local train stationary at a platform, and, as the carriages overturned and piled up, a down Liverpool express hauled by two locomotives crashed into the wreckage; both these locomotives ricocheted off the crushed carriages and over the down platform. Three passenger carriages in the local train, three in the Perth and seven in the Liverpool were crushed; an eight-wheel milk van, a kitchen car and two parcels vans were also smashed to bits. Thirteen eight-wheeled vehicles, normally about 720ft long, were compressed into a space 135ft long, 54ft wide and 30ft high. The wreckage was piled up onto the road above the station. The up Watford electric line, up and down fast lines and the down slow line were blocked; the up slow was not obstructed. Just 21¼ hours after the accident, the down slow was brought back into use, giving a double-track route past the crash site. The station footbridge had to be removed to extricate the engine of the Perth train, and at 8pm on the fourth day after the accident, normal working was resumed on all lines, although with a speed restriction on the up and down fast lines. The station footbridge had also been restored within that time.

On 7 January 1966 a down freight train running at 45mph was derailed at Uffington, Oxfordshire, owing to

Once a familiar sight awaiting their next duty at numerous locomotive depots, rail-borne cranes could be used either for emergencies or for civil-engineering projects, particularly where new bridges needed to be erected. On 4 June 1982 Class 31 No 31138 passes Stratford with the local shed's breakdown train. *Mark Symonds*

an overheated axle bearing which then sheared off, bringing the wagon down onto the track. Had there been signalmen to see that the bearing was getting hot the derailment would not have occurred. The derailed train tore up ¾ mile of track before coming to a stand. The cranes were ordered out and arrived in quick time, rather like the fire brigade. The damaged wagons were lifted clear of the line and put down the bank. The track was then relaid, and within 24 hours normal working resumed (with a speed restriction, of course) on new track.

Rail-mounted cranes of various lifting capacities were stationed at major locomotive depots along the line of route, and these were rushed at express train speed to any derailment; they were also used to lift track sections during track re-laying. In BR days one Region would, if requested, lend a crane to another where an accident happened close to a Regional boundary. Even in the days of the private railway companies, this is known to have happened. When there was a derailment at Stoke Canon in 1938, the Southern Railway at Exmouth Junction shed volunteered (without being asked) to send — and did send — a crane to assist the GWR's cranes. As with a disaster at sea, everyone rushes to assist and 'private property' ceases to be relevant.

We are so much cleverer nowadays and do not incur costs that can be avoided, and to that end we have scrapped most (but not quite all) of the great cranes, purpose-built for railway work, and instead hire in road

Even on the best-regulated railways, things can and do occasionally go wrong, and the availability of equipment such as cranes is essential if the line is to be restored as soon as possible to operational condition. In April 1965 two steam cranes attended a derailed freight train near Cumnock, on the Scottish Region. *W. J. V. Anderson*

cranes when occasion requires. On 28 February 2001 a Land Rover drove off the M62 motorway at Great Heck, near Selby, and onto the East Coast main line in front of an up express train. The train was derailed and was then foul of the down line and came into collision with a down goods train. There were no railway cranes, even though York was 20 miles to the north and Doncaster 11 miles to the south — the former once one of the greatest railway centres in the land and the latter not only a huge junction and motive power depot but also a major railway factory. In order to clear the line a road had to be constructed down to the site to enable a road crane to be brought in, and the task of clearing the line took 10 days. During this time buses ferried passengers between York and Doncaster; to compound the low state of efficiency to which Britain's railways had sunk, trains leaving Doncaster for the south were not always held to connect with the bus coming down from York when road congestion delayed that bus and made it late at Doncaster.

Railtrack divided the country into 'North' and 'South';

the southern half scrapped all its conventional jib cranes, but the northern half kept a few. The scrapped cranes and associated equipment — lifting beams, chains etc — were purchased by a scrap dealer who, without actually seeing the items, sold them on to preserved railways, and the residue was cut up. In August 2002 Anglia Railways sold its entire stock of vehicle-recovery and re-railing equipment (such as it was) to Freightliner; Anglia says it is not its business to re-rail trains, since it does not own the trains or the track.

Disposal of the jib cranes left large depots such as Stratford, Ipswich and Norwich without any heavy-lifting equipment, the nearest being at Old Oak Common in West London. The men to work these machines will come from all over the country — there is no such thing as a local gang. Men being called out to a derailment are quite likely to say: 'Is it a big job? What's the rate? Is it worth me turning out?'

In the days when there were cranes and recognised, close-knit breakdown gangs at major depots, these teams knew each other and worked together. These men were trained mechanics, fitters — perfect for breakdown work.

On 20 May 1975 a train from the London Brick Co's Calvert Works was derailed at Claydon LNE Junction. This view, taken on the following day, shows the Willesden Junction crane assisting in the recovery of the derailed wagons. With the demise of much traditional freight — including brick traffic such as this — the incidence of freight-train derailments has declined, but they can and do still occur. One wonders where the nearest rail-borne crane would be if a similar incident occurred at Claydon today. *Kevin Lane*

They were a group, a platoon. They were like the crew of a fire engine or an ambulance. Apart from the obvious advantages of this in terms of the speed of response to an incident and the speed with which an incident was cleared up, there was the added advantage that a recognised team of men — a coherent, trained group — could and did take part in exercises with the emergency services. They could and did practise working together. There was a 'cohesive structure', an 'organised plan', a 'team' — there were all those things which modern management is so fond of spouting on about and yet does not possess.

Heathrow Tunnel

The railway line from Paddington to Heathrow Airport turns south off the Great Western main line at Airport Junction, near West Drayton, about 12 miles from Paddington. It runs in a 'cut and cover' tunnel and just north of the M4 enters a true tunnel in which it runs under the Central Area and thence to Terminal 4.

The proposal for the new railway was approved in 1988. The multiplicity of companies involved in its design and construction would have astonished I. K. Brunel, who had designed and supervised with one assistant, and the number of participants must surely have added to the complexities of building it. Ultimately responsible was the British Airports Authority (BAA), through its wholly owned subsidiary Heathrow Airport Ltd (HAL). HAL duly set up an Engineering Projects Group, whose staff were transferred in 1991 to BAA Group Technical Services Division and worked thereafter under the supervision of Taylor Woodrow Management Contracting Ltd, taken on by BAA as the Project Managers. Other advisers, consultants and designers employed for certain jobs came and went, but Mott MacDonald, the principal design agency, was retained to assist BAA. Mott MacDonald in turn was instructed by BAA to engage a firm of specialist consultants — the Dr Sauer Company — to oversee construction of a trial tunnel.

Construction was to be by means of the New Austrian Tunnelling Method (NATM), whereby the tunnel, once excavated, is quickly supported by steel ribs; a concrete lining 250mm thick is then sprayed onto the tunnel to form a concrete tube. Clearly this has the potential to be a relatively cheap way of making a tunnel — if all goes well. However, there had been 39 major collapses of NATM-built tunnels worldwide, and there was no previous experience in the use of this technique in the prevailing London clay.

London clay is easy to excavate because it is soft and wet, but the cavity formed in the clay is subject to vertical and horizontal pressures from the surrounding earth and, if not quickly supported, will inevitably collapse. Tunnelling through the clay cuts through and disturbs the numerous geological 'planes of weakness', which are thus rendered unstable.

In 1990 an investigation into the clays to be tunnelled was undertaken by HAL. The methods used were conventional, and the survey was of a general nature and did not take into account the particular needs of the NATM system of tunnelling. A single-bore, 100m-long trial NATM tunnel was then driven, by a contractor who played no further part in proceedings. The operational tunnel would be triple-bore, entailing the removal of a great deal more clay.

On 7 March 1994 Balfour Beatty was awarded the contract to design and build the triple tunnel, to accommodate an up line, down line and concourse tunnel and underground station. This contract permitted Balfour Beatty to certify the quality of its own work; I. K. Brunel might have found something there to criticise. Balfour Beatty then engaged Mott MacDonald to predict ground settlement over and around the tunnels, which would pass beneath the Underground's Piccadilly Line and the Camborne House office building on the surface of the site. As this involved Balfour Beatty in expense, there might have been a conflict of interest, since Mott MacDonald was also working for HAL/BAA, but so-called 'Chinese walls' were cleverly set up to prevent any such conflict.

The NATM system of tunnelling required 'primary' and 'secondary' linings; the contractor's staff designed the former and Mott MacDonald, working for HAL, designed the latter.

Balfour Beatty placed a sub-contract with the Austrian inventor of the tunnelling system — Geoconsult — to design the tunnels and provide on-site supervision. Geoconsult considered that the minimum size of supervision team to be one senior engineer and two foremen, but Balfour Beatty thought one senior engineer was enough and refused to pay for more. The result was that Geoconsult's site engineer was present only during the day, Monday to Friday, and was not there at weekends — but the work went on 24 hours a day, seven days a week. Balfour Beatty appointed three of its own men as junior engineers to work under the Geoconsult senior engineer. These men had limited experience of tunnelling and none at all of NATM. They were called 'NATM engineers' but in reality were there merely to install monitoring equipment and take readings.

The HAL/Balfour Beatty contract did not specify any factors of safety in the design of the tunnel linings, so Geoconsult adopted its own, but these did not take into account the load that would be placed on the surrounding clay when cement grouting was pumped into it. Rigid, semi-circular lattice girders were specified for insertion into the roof, but no such support was given to the lower half of the excavation. Such girders also provide a datum point from which to gauge the thickness of the sprayed concrete, and the absence of girders in the lower half made it difficult to maintain a constant thickness in the concrete lining.

The Geoconsult engineer was partially integrated into the Balfour Beatty management team but did not attend all the engineering meetings between HAL and Balfour Beatty and was not on BAA's circulation list of minutes of meetings. Balfour Beatty developed risk assessments without consulting Geoconsult, its technical adviser, and these assessments did not fully identify risks to tunnellers. No risk assessment was carried out with regard to repairs to the tunnel during construction, so that work that involved major risks was undertaken without adequate safety provision. A site safety plan was developed, but this did not provide guidance on how to prevent the tunnel from collapsing.

The original contract between Geoconsult and Balfour Beatty made the former responsible for the provision of computer software to process tunnel-monitoring data, but Balfour Beatty struck out that clause and provided its own software. It was considered good practice to validate such software with the Association of Geotechnical Specialists, but Balfour Beatty did not do this. The Geoconsult site engineer was not impressed and wrote to Balfour Beatty, asking it to change to his recommended software and surveying systems, but this request was rejected.

The lack of any independent quality control militated against quality. Having allowed Balfour Beatty to certify its own work, HAL expressed its own concerns on 20 July 1994. Repetitive construction defects occurred in the Central Area pedestrian concourse tunnel, which was unsound from the outset. Balfour Beatty and Geoconsult were aware of defective workmanship but appeared not to realise the significance of that knowledge. By the time defects had been noted by the HAL team and a request for corrective action reached Balfour Beatty, work on the tunnel had advanced beyond the defective part, requiring the newly applied — and perfect — lining to be cut out so as to get back to the defective section for remedial action to be taken. The need for such remedial action put the project behind schedule, meaning that the outstanding work had to be accelerated in order to catch up.

On 15 July 1994 Geoconsult's site engineer wrote to Balfour Beatty's engineering manager, pointing out — along with many other concerns — the poor workmanship in the construction of the tunnel's lining. Balfour Beatty did not accept any of these criticisms.

The approach to the infamous tunnel at Heathrow. *Brian Morrison*

During the first half of July the triple tunnels had passed below Camborne House. The ground at once began to sink, and the building with it. Geoconsult had earlier done a finite-element analysis and predicted that the ground above the tunnels would sink by 9mm — insufficient to cause structural damage to the building. Yet subsidence above the single trial tunnel had been between 25 and 35mm. By mid-July the ground under Camborne House had subsided by as much as 22mm and was still sinking. On 22 July BAA/HAL raised the matter with Balfour Beatty, which said that it was due to circumstances beyond its control.

With Camborne House continuing to sink, Balfour Beatty was faced with a difficult decision: it could let events take their course and pay for repairs to the building once the ground had stabilised, or it could stop the subsidence before any damage was caused by using 'grout jacking', whereby cement would be pumped into the ground below Camborne House and thus force the ground upwards. It opted for the latter course, but Balfour Beatty's engineer overseeing this 'compensation grouting' had no experience of the process. Pipelines were nevertheless installed, and pumping of cement began on 5 August, leading to an immediate, stabilising effect on Camborne House. However, it is a basic law of nature that any force applied in one direction has an equal reaction in the opposite direction — so jacking something up means exerting a force downwards. What would happen to the ground below the grout?

No-one, apparently, thought of arranging communication between those pumping cement into the ground and those monitoring the shape of the concourse tunnel below, which started to become distorted as its (designed) flattened base (or 'invert') took the strain of the load placed on the crown of the arch. The three Balfour Beatty 'NATM engineers' did not detect this from their instruments.

Between 5 and 15 August 29,770 litres — 6,615 gallons — of cement was forced into the ground below Camborne House, immediately above the concourse tunnel. The ground lifted unevenly: the southern corner of Camborne House returned to its original level, but the northern and northwestern corners rose only by 30%. Meanwhile, below ground, the tunnel crown had been forced downwards 50mm for a distance of 35m on each side of Camborne House. On 16 August cracks were noted around the circumference of the tunnel lining, while the somewhat flattened invert of the tunnel was cracked longitudinally, with one edge riding up over the other.

The concourse tunnel was inspected for damage under Camborne House, 54m from the tunnel entrance, and also northwards to the tunnelling face, 104m from the entrance; no inspection was made southwards of 54m.

Repairs began on 23 August. There was, in fact, a major fracture of the base or invert of the tunnel lining going south from the 54m point, below Camborne House, but the full extent of this crack was not investigated. One of the 'NATM engineers' did write a note in the Engineers' daily-occurrence book, asking whether repair work was to stop at 54m, and the Geoconsult site engineer wrote alongside this 'NO!' There was then a failure of communication. The Geoconsult man, believing his two-letter word sufficient to inspire the Balfour Beatty men to continue with their investigations and repairs, said or wrote no more, but, because he did not raise the subject further, the Balfour Beatty men thought they did not have to take further action. On 9 September the Geoconsult engineer reported to Balfour Beatty that the situation was repaired and under control — which it was not.

It was further discovered during the repair work that the thickness of the concrete lining was, in places, only 50mm instead of the correct 300mm.

On 2 September BAA asked Balfour Beatty for an assurance that the concourse tunnel was sound. On 20 September Balfour Beatty gave that assurance, although evidence showed the tunnel was not sound. On 16 September, meanwhile, Geoconsult wrote to Balfour Beatty, suggesting alternative management procedures, better processing and presentation of tunnel-monitoring data — and the need for better software to accomplish this. On 26 September the Concourse Tunnel Construction Superintendent, employed by BAA, challenged Balfour Beatty on the integrity of the tunnel, unaware that senior BAA management had already done this; duly informed that BAA had accepted Balfour Beatty's assurances, he reluctantly formed the view that there was nothing more he could do.

The unrepaired section of the concourse tunnel, southwards from 54m to 0m, continued to distort, and Camborne House sank 40mm. It was not until 4pm on 18 September that Balfour Beatty stopped pumping cement into the ground above the tunnel beneath Camborne House, whereupon the building settled by 9mm. Between 21 and 23 September almost 3,000 litres of cement was pumped in under the building, but by 10 October it had sunk another 7mm; below, the concourse tunnel was gradually giving way. By 2 October no fewer than 117 requests for corrective action for this area had been raised by BAA with Balfour Beatty. A disaster was imminent, yet apparently no-one in authority saw it coming.

Geoconsult did not consider as significant the fact that Camborne House had sunk 47mm. On the surface, in line with the tunnel below, there was a depression between 35 and 55mm deep. On 14 October the Geoconsult engineer noted accelerating movement in the concourse tunnel, at the southern end of the site. He decided to go home for the weekend and assess the data collected. When he returned on the 17th he saw that the concourse tunnel was still moving and decided to speak to his superior the following day when he arrived from Salzburg. Balfour Beatty's site engineer inspected the same tunnel on 17 or 18 October but noted nothing untoward.

On 20 October the Balfour Beatty 'NATM engineer' on duty since 7am went home sick at 10am. The Project Director was already off sick, and the Construction Superintendent had just returned from sick leave. He did not go down into the tunnels because he attended the morning meeting with BAA officials, and in the afternoon he was again ill. The meeting he attended was routine, between HAL/BAA and BB engineers, to consider progress with design work. No-one mentioned the movements in the tunnels or the matter of repairs. It was as if they were all becoming used to seeing the tunnel lining filled with cracks and a large building on the surface sinking into a steadily forming depression.

At around 6pm on 20 October, concrete began to spall away from the linings and fresh movement was detected in sections of the lining that had recently been repaired. Vertical cracks were appearing. Into the night, bits of concrete were bursting off the lining, and cracks were growing visibly along the length of the tunnel. BAA and Balfour Beatty engineers came and photographed large cracks in the concrete walls but did not realise what it portended. At 11.45pm the Balfour Beatty 'NATM engineer' noted that the crown of the down-line tunnel dropped 25mm in an hour; the inexperienced supervisors did not evacuate the workers, although they were sufficiently concerned to call in the Site Agent.

The tunnel was now cracking up rapidly. Plates of concrete were splitting away from the walls and cracks were appearing, old cracks growing. It was an ordinary tunneller who, at midnight, told his pit boss that they ought to get out, yet it was 12.45 before the Site Agent, Tunnel Superintendent and Geoconsult's man decided that the situation was irretrievable. The men went to the surface up the fuel-depot shaft to muster for roll-call in the staff canteen in Camborne House. The ground was moving as the men walked to the canteen, and, shortly after they arrived, the canteen began to tilt over, causing them to leave hurriedly. The tunnels finally collapsed at around 1.15am, creating a deep depression in ground that crowds of workers had just crossed and taking Camborne House into the abyss.

On 15 February 1999 at the Old Bailey, Balfour Beatty and Geoconsult were found guilty of breaching the Health & Safety at Work Act 1974. Balfour Beatty was fined £1.2 million and Geoconsult £500,000. Each was ordered to pay £100,000 costs. The judge said: 'The collapse was one of the worst civil engineering disasters in the UK in the last 25 years.'

Construction Calamities

In the early 1970s, after almost two centuries of discussion, work actually started on the construction of a fixed link between Britain and France. As work progressed, the new bores gradually extended for some 400m under the English Channel before the new Labour Government led by Harold Wilson decided to pull the plug on the venture and work stopped. It would be a further two decades before the fixed link was finally completed. This view shows the aborted bore in Shakespeare Cliff, near Folkestone. *Monitor*

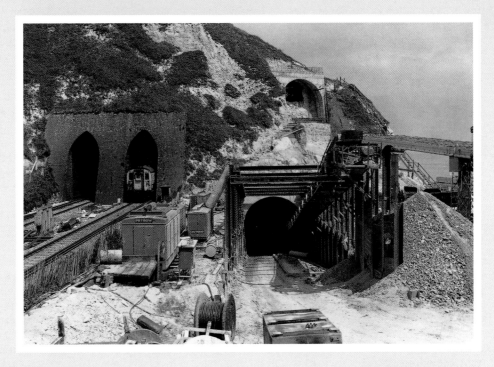

In March 1979, during work to increase the loading gauge of Penmanshiel Tunnel to take 8ft containers, the tunnel roof collapsed, tragically killing two of the workmen involved. The East Coast main line was blocked, and the decision was taken to realign the route at this point rather than try and reopen the original tunnel. Four months later, on 22 July, work on the new alignment was well advanced, as this view, showing the north portal of the abandoned tunnel and the new trackbed, illustrates. *G. A. Watt*

Heathrow Express

The Heathrow Express runs over a brand-new route from Airport Junction (near West Drayton on the Paddington–Slough line) to Heathrow, the tracks and the curves having all been built especially for the electric units that would run over it. The 15min-interval public service commenced on 19 January 1998 but was suspended on the 31st, the wheels of the new trains having been so badly damaged that they had all to be withdrawn for repairs. The purpose-built track had been laid with curves too sharp at Airport Junction and Heathrow Junction, so that the flanges were being subjected to excessive wear — sufficient to wear them dangerously out of shape in a mere 12 days! The wheels were duly removed for re-profiling; by 3 February there were enough serviceable units to run a half-hourly service, and by 14 February all were back at work. However, the wheel-lubricators in the track had not yet been corrected, so every night the entire fleet of EMUs had to be turned around so as to spread the wear over all wheels and thus double the length of the time wheels would last before the units all had to be taken out of service and again re-profiled. The units were turned on the non-electrified Greenford triangle, hauled by a Class 37/6 diesel — a time-consuming and expensive process.

Flanges squealing, as a result of the sharp curve and inadequate lubrication, Heathrow Express unit No 332009 comes off the Great Western main line and into the temporary station at Heathrow Junction on 19 January 1998. *Brian Morrison*

Great Yarmouth Derailment

On 25 August 2001 the 10.05am Great Yarmouth–Liverpool Street was derailed as it pulled out of Great Yarmouth at walking speed. The train consisted of 10 corridor coaches and was hauled by Class 47 diesel locomotive No 47787, which carried the name *Victim Support*. The victims of this fiasco were the passengers.

The train was routed from Platform 2 on to the Reedham line (Line 3). As the train passed over the points at the outer end of Platform 3, the rear axle of the locomotive and all wheels on the leading carriage came off the rails. The driver brought the train promptly to a stand. They were 'on Olde England', but the train was upright and intact. The two rearmost carriages were still alongside the platform, so the passengers could easily disembark, and after considerable delay they were taken on to Norwich by bus.

In the 'bad old days' of British Rail, the Great Yarmouth signalman would have telephoned the foreman at the locomotive shed and requested the breakdown vans; these would have been at the scene within 30 minutes, and the train would have been re-railed in an hour or so. This is not wishful thinking but was normal practice on the steam-hauled railway.

However, this was the modern railway, where the culture of 'monolithic' BR has been swept away and replaced by the discipline of the market place. Of course, this 'discipline' was designed not for getting trains re-railed but to make a profit. So first of all there had to be an agreement as to who 'owned' the incident — who was to blame — and therefore who would pay for the damage. The wooden sleepers were in such a state of decay that they could have been broken up and carried away on spades. The 'chairs' holding the rails had, quite clearly, been forced apart by the wheels' passing over them, because the sleepers were so rotten as to allow the bolts to move through the wood. However, the contractor responsible for maintenance of the track, Balfour Beatty, declined to accept the blame, as did Railtrack — since one or other would then have to pay for the rescue operation. As blame was not accepted by the obvious culprits, the arbitrator — AEA Technology of Derby — had to be called in. But this outfit had only one set of staff capable of adjudicating on the matter, and they were in Glasgow, dealing with another derailment.

So there sat the train, on the new, exciting, efficient, disciplined, free-enterprise railway — on the ground. Dead. Great Yarmouth station was reduced to a single platform (1), with consequential delays.

Not until 1.30pm did the various officials nowadays required to attend a minor derailment begin to arrive. First came the man from HM Railway Inspectorate. He got there 15 minutes before the On-Call Rail Manager arrived. The train had derailed $3\frac{1}{2}$ hours earlier! British Transport Police sent a Scene of Crimes Officer from Euston, and he joined the crowd of people staring helplessly at the stranded train. The real crime that had been committed was to sell the railway to persons ignorant of railways but expert in 'cultures' and 'market places', but, of course, that was not a crime but the law. At 6pm yet more railway officials arrived to stand and watch, but no-one could do anything to clear the line until 7.03pm, when the AEA Technology people arrived. It took only a few minutes for them to tell everyone what was obvious — the track was defective. Work began on re-railing at 7.30, and by 8.10pm the engine was back on the rails. At 9.30 the carriage was back on the rails. It had taken the wonderfully systematised railway 10 hours to complete what the old railway, run by horny-handed sons of toil, would have done in an hour. And this is only one of any number of similar inefficiencies which regularly occur for the same reasons.

Virgin
and the 'Voyagers'

Sir Richard Branson, one of the greatest exponents of free enterprise in Britain, took over the franchise for cross-country trains in January 1997 and the WCML franchise in March 1997. The nation was told to look forward to a bright new railway future enlightened by the great man's 'vision'. Sir Richard promised that trains would be 'fun', that he would provide hitherto unknown luxury and that all this would be accompanied by lower fares. Sir Richard had never been a railway manager, but this was seen by other non-railway persons as a jolly good thing.

Running on the WCML were freight trains and 'outer commuter' trains, all of which had an equal right with Sir Richard's trains to be there. Of course, the primary blunder was in creating a situation where the traffic flows on the busiest trunk route in Britain were divided between separate businesses, each jockeying for its own advantage. Sir Richard is a very good jockey and was soon several lengths ahead of the others in the race. This, however, had nothing to do with running a railway, where tens of thousands of people every day had top be moved north and south, east and west. Sir Richard wants to have exclusive use of the up and down fast lines of the WCML and to consign everyone else to the slow lines. From a railway-operating point of view this is a nonsense, and whether it will come to pass remains to be seen.

In January 2000 the Midland Rail Users' Association spokesman, Philip Davies, said: 'Passengers are paying through the nose for a service that does not meet their legitimate expectations.' Under the much-denigrated British Rail the Euston–Birmingham expresses paid their way and in some years even made a modest profit. It was the success of the BR Inter-City services that caused the privatisers to think that they ought to be taking the profits. In the first three years (1997-2000) of the Virgin franchise the price of a Standard-class return ticket from Stafford to Euston rose by 46% to £99.50. There was no increase in luxury, the trains were the same, but the standards of punctuality were lower. A return ticket from Manchester to Euston cost £130 by Virgin Trains, but a traveller could get a £30 return if he or she booked it several days in advance.

A passenger who often booked in advance for a ticket on the 05.20 Manchester–Euston found he was unable to buy one in September 1999. He wrote to Virgin Trains to ask why, and after a delay of three months he received the explanation: 'We have recently taken the decision to remove all advance purchase fare availability from morning peak trains. This is because some business travellers had been booking the cheap tickets in advance.' The inference was that booking in advance was a rather shady subterfuge to deprive Sir Richard of his just deserts. The result of this ruling was that the early-morning train was regularly all but empty leaving Manchester. Even in January 2000, Virgin was still letting it leave with 15 people on board rather than re-institute the £30 ticket. What is the point of running a railway on the basis of spite?

Getting luxury back into trains must by definition be an expensive operation. In days past there was great luxury on railways, but the operational/engineering side was always to the fore. Pullman cars could still be coupled to ordinary carriages, the buffers were the same height, the couplings were universal, and even a small tank engine could haul them should that be necessary. Luxury is not to be achieved by painting in bright colours non-standard vehicles and attempting to run them with the minimum of maintenance staff. Rail luxury is brought about by the whole system, not just by PR hype. There must be speed and reliability, brought about by good organisation, simplicity of design and good maintenance.

Reliability in the past was very good because of the robust mechanics of trains. The ancillaries of the train were in keeping with the rugged service the train had to 'deliver'. Simplicity was a noticeable feature. This is no longer the case where Sir Richard's new trains are concerned. In this era of vastly increased passenger rail travel, an ordinary mortal would expect that these trains can carry more people. But ordinary mortals do not run the Virgin franchise. The Class 220 'Voyagers' are four-car trains, but — and here's the clever bit that no ordinary railwayman would think of — they run at double the frequency of the old eight-car trains. Your average railway manager would see that this doubles the number of trains on the line and will lead to severe congestion at terminals

One of the longest Virgin CrossCountry services operated during the summer of 2000 was that between Ramsgate and Glasgow Central. On 27 May of that year Class 47/4 No 47845 *County of Kent* passes Bickley Junction with the 12.10 Ramsgate–Glasgow. *Brian Morrison*

— and therefore would not entertain the idea. But permission for this wasteful use of space came from the Office of the Strategic Rail Authority. This is the advantage of not having a railwayman at the helm: entrepreneurial — and amateur — railway operators are not 'hidebound' by an 'old-fashioned culture of British Rail' If the line is overcrowded with unnecessary trains, 'sports car' acceleration ceases to be of much importance.

An eight-car HST would have two First-class cars carrying between them 96 passengers, a restaurant car with seating for 24, four Standard-class cars carrying a total of 288 and a Standard-class/brake van seating 31 — an overall total of 415 passengers, not including the 24 seats in the restaurant car. A four-car HST, formed of one First-class, two Standard-class and a brake van would carry 223. The new Class 220 units, built for an age of greater demand, have four cars and carry 188 people.

These great new trains are controlled largely by a computer which appears to reside on the roof. The alleged 'sports car' acceleration thrill is frequently dampened by the long delays inflicted by malfunctioning silicon. On 9, 10 and 11 October 2002 five 'Voyagers' travelling along the sea wall at Dawlish were brought to a stand when the rooftop electrical equipment was short-circuited by sea water breaking over the sea wall and landing on the roof. Of course, manufacturer Bombardier complained of unusually high tides and of Virgin's allowing the trains to run when there were such big waves.

Right: A pair of 'Voyagers' at Birmingham International on 23 July 2002. On the left is No 220025 *Black Country Voyager*, forming the 06.20 Poole–Newcastle, while on the right No 221126 *Captain Robert Scott* waits to depart with the 10.46 Birmingham–Edinburgh. *Brian Morrison*

Below: On 30 July 2001 Mk3 DVT No 82124 *The Girls' Brigade* heads Virgin Trains' 09.45 Liverpool Lime Street–Euston service through Tamworth Low Level. Power was provided by Class 86/2 No 86233 *Laurence Olivier. Brian Morrison*

But trains that can run only in ordinary conditions are not really worth putting on rails. Other trains ran through the waves during that period without stopping, but then they did not have electrical equipment on their roofs.

On 4 July 2002 the 14.36 Manchester–Birmingham was formed of Virgin Class 221 'Super Voyager' DMU No 221109 *Marco Polo*. It had about 200 people on board. At 16.15 at Penkridge, between Stafford and Wolverhampton, *Marco Polo* was brought to a sharp stop by an emergency brake application, made not by the driver but by the on-board computer. The driver was unable to get

the computer to let go of the brakes, so there stood *Marco Polo*, voyaging nowhere. The 15.10 Liverpool–Poole, another 'Voyager', was brought up and coupled to the Manchester to see if its computer could persuade the other's to let go of the brakes. The effort proved too much for the second train, and it was declared a failure at 17.50. Virgin Trains took an hour to despatch a Class 86 electric locomotive from Rugby, carrying an adaptor coupling to tow the trains. The rescue engine arrived on site at 20.45 but could not couple then up. Even when skilled mechanics from Bombardier — builder of the 'Voyagers'

— arrived, the coupling could not be achieved. This was because the locomotive was a conventional railway vehicle, while the 'Voyager' was the flower of the new, entrepreneurial railway, which declines to be coupled to anything as socialistic as a locomotive; it is the railway version of 'I wouldn't touch it with a bargepole'. It seems like a waste of time to send out a conventional locomotive without the necessary 'barge pole' in the form of a barrier wagon, with conventional couplings at one end and a free-enterprise 'Dellner' coupler at the other.

The word 'passengers' must not be used — it does not convey sufficient idea of their status — for they are 'customers' and very nearly sacred for that reason. Being given the status of 'customer' is so much more assertive than being merely passive 'passengers'. According to the *Daily Telegraph* for 5 July 2002, the customers sat assertively in Sir Richard's luxury train for five hours without toilets — the flushing mechanism is electric, and there was no power — and without air-conditioning. There being no way of getting fresh air into the carriages, the atmosphere became sweltering. At the end of the five hours they were 'de-trained' and asked to walk along the side of the line to the Liverpool train, which had air-conditioning and toilets. Two trainloads of passengers sat — or stood — for another three hours until the heavily delayed 09.15 Aberdeen–Plymouth HST drew up alongside at around 23.40, presumably running up the down line. The stranded customers were then escorted across the tracks on 'safety boards' — one must be very careful with one's customers — and helped onto the waiting train. The Aberdeen reached Birmingham at 00.30 and terminated at Bristol at 02.57, nearly 8¼ hours late.

Virgin refunded all fares in full, paid for people to have taxis or paid for hotel accommodation. Jim Rowe, Virgin Trains' Corporate Affairs Manager — who had not been on the train — said: 'We can only apologise — it was a bit of a nightmare.' The last seems to be a masterly under-statement. But what can he do? He has a train that seems to come from another planet and won't communicate with any other train, he owns no track, no locomotives — he has very little power over anything, really. All he can do is apologise and try to find someone else to blame. Of course, on the fragmented railway there usually *is* someone else. Chris Green, Virgin Chief Executive, said that the 'Voyagers' have 'sophisticated software that is prone to over-react'. So it is not *that* sophisticated. 'Sometimes,' he said, 'bounce from poor track [so there's where the problem really lies] makes the computer believe that the tilt system is breaking loose and so makes the train stop. We are modifying the software.' It might be a good idea to allow the drivers to drive the train, but probably they are not sufficiently sophisticated. Mr Lovett, also of Virgin Trains, blamed a bird for getting in the way of the train and damaging the roof-mounted electrical equipment. Between bumpy tracks and bird-strikes there has been a fair number of computer-generated, un-releasable brake applications on the Virgin 'Voyagers'. On 8 July Nos 221119 *Amelia Earheart* and 221122 *Dr Who* failed at Lancaster with the 06.50 Edinburgh–Bournemouth; after 75 minutes the pair were finally separated, No 221122 being left behind, dead. On 12 July No 221102 *John Cabot*, coupled to No 221107 *Sir Martin Frobisher*, failed for the same reason near Didcot, but after 64 minutes the driver was able to persuade the computer to stop messing around, and the train got on its way. The mention of *Sir Martin Frobisher* near Didcot brings to my mind the fact that there was another engine of that name, Maunsell 'Lord Nelson' No 30864, which used regularly to pass Didcot station. I see the great engine and its 14 coaches in my mind's eye, and then the full pathetic amatuerishness of allowing computers rather than men to control trains becomes well-nigh unbearable.

The Class 221 'Super Voyagers' run as five-car sets but *pro rata* carry no more people than do the 'Voyagers'. They cannot help but be overcrowded because they are too small for the large numbers of people wanting to travel. The privatisers have actually boasted that it is due to their great efforts that many more people want to use the railway — and then they build trains that are too small to cope. That is truly magnificent. Overcrowding has even led Virgin to instruct its staff to tell lies to the sacred customers. In the early part of 2003 an over-crowded Virgin 'Voyager' left Oxford for Birmingham, but customers were walking about looking for the seats they had reserved but could not find. Shortly the Senior Conductor made an announcement over the tannoy, apologising for the lack of seats and stating that this was due to a carriage having been removed because of a defect. This was untrue — the train was a Class 220 'Voyager' still in full possession of its four cars. Nor was this an isolated incident.

The technical problems of the trains reduced the size of the available fleet, and in January 2003 the problem became so acute that Virgin was forced to retain in service the solid, roomy, British Rail-designed HST. Why not build new trains like these willing workhorses?

So, in June 2003 how much fun are Sir Richard Branson's trains? How much better are they than those run by BR? Are they worth the huge subsidies he has taken from the public commonwealth? In 2003 he was due to pay £8.2 million to the Government as a share of the profits he had made from these trains. Instead he is being paid £282 million on top of the £106 million he was paid by the taxpayer last year. I feel bound to ask whether we can afford private enterprise.

The East London Extension

In 1988 London Underground Ltd (LUL) published its East London Assessment Study. The proposal was to create a cross-London route by a combination of rebuilding old rights of way and refurbishing existing routes to link north–south across London and across the Thames. The need had been identified by the Buchanan Report as long ago as 1963.

The project consisted of joining the existing East London Railway to the North London Railway, starting in Kent at the southern end of the East London line and running from the BR suburban lines at New Cross and New Cross Gate, under the Thames to Whitechapel. In addition, it was intended to utilise a derelict trackbed from the East London line at Surrey Docks to Peckham. North of Whitechapel the track would be raised over the cutting through which runs the East Anglian main line into Liverpool Street — requiring the removal of part of a derelict viaduct, built between 1837 and 1840 and disused since 1874 — and then on a new viaduct across Shoreditch High Street onto the disused viaduct of the old North London line near Dunloe Street. This viaduct had once carried four tracks from the Broad Street terminus — now demolished and redeveloped as Broadgate — north to Dalston Junction, where a forked junction connected, east and west, with the North London line. The latter is a most excellent railway, which the Government of the 1980s would have closed had it not been for the strenuous opposition of the Leader of the Greater London Council. Refurbishing the existing viaduct between Shoreditch and Dalston and relaying the track, along with the construction of a new viaduct over Shoreditch High Street, would enable the provision of new commuter services from Surrey and Hertfordshire to Southend and London's Docklands. In addition, long-distance trains from the West Country and from the West Coast, Midland and East Coast main lines would be able to use the line into Kent. It was a most enterprising — and highly necessary — scheme, giving those who normally only worship British railway engineering if it is 160 years old something modern to praise.

The East London project is a brave and imaginative programme which will make the railways more useful and reduce road traffic congestion. Reducing the number of cars in London will reduce noise, air pollution and respiratory disease. It is fully deserving of support, but still it is deemed to be 'not self-funding' — by some arcane and devious method designed to prevent railways from being built. As ever, an arbitrary and theoretical 'accountancy' stood in the way of a practical solution towards a cleaner environment and greater convenience for ordinary people. However, the dedicated band at LUL pressed on with the project to provide a cleaner, quieter, less polluted London.

London Underground was to provide the money for the first stage of the great work — £7.5 million to make safe the brick walls of the ancient Thames Tunnel. The bricks had been laid between 1828 (at the south end) and 1843 (at the north end) and were showing their age. Water, under pressure from the surrounding earth and the river only just overhead, hissed through the bricks, and where this happened the ingress had been stopped by drilling a hole and hammering in a wooden peg bearing a strong resemblance to a piece of broom handle. The method chosen by LUL's engineers was to coat the brick walls with concrete — 'shot-creting'.

When the organisation known as English Heritage was informed of these preparations it made representations to Stephen Dorrell, then Heritage Secretary in the revolving door of politics, to have the tunnel 'listed' as a structure of historic importance which was therefore not to be altered. The irony of the English Heritage attitude was that the object of its admiration, being inside an underwater tunnel, was invisible to everyone except the driver of the train, and, if the work were not carried out, the tunnel would fall down and cease to be an object of interest. Mr Dorrell rejected the English Heritage request to give the tunnel 'listed' status, and the 'shot-creting' work was scheduled to begin on 25 March 1995. The train service through the tunnel was to be suspended for six months, during which time replacement buses, running via Tower Bridge, would be provided.

All the arrangements were in place when, at 5pm on 24 March, Stephen Dorrell suddenly changed his mind and agreed with the English Heritage application to have the tunnel registered as a 'listed building'. The tunnel remained closed to trains, leaving the wheels of bureaucracy set in motion by English Heritage to grind interminably over the carefully planned project.

London Underground now had to submit an application to the Docklands Development Corporation for permission to carry out the work that would keep the tunnel standing up, 'listed' status being insufficient for that purpose on its own. English Heritage went off to find 'a panel of experts' to find a better way of saving the tunnel than that selected by LUL's engineers. The outcome of the 'listing' was highly ironic: the work was delayed three years, creating huge additional costs for LUL (which, of course, English Heritage did not have to pay), at the end of which the panel of experts brought in by English Heritage decided that 'shot-creting' was indeed the only way to save the tunnel. To save the face of English Heritage, the ultimate irony: four 'panels' of the arches had to be retained in their original condition after some serious repairs were carried out, again at great extra cost to LUL — and these panels can only be seen properly by the drivers of the trains.

Standing in the way of the extension northwards of the East London line are the derelict remains of Bishopsgate Viaduct. This once carried the Eastern Counties Railway, later known as the Great Eastern Railway, to its Bishopsgate terminus on Shoreditch High Street. English Heritage refers to it as the 'Braithwaite viaduct' after its engineer, the incompetent John Braithwaite. He it was who brought the ECR into Bishopsgate, but, instead of following the falling ground, he maintained his level, so that the tracks came into the terminus on a $1^1/_4$-mile-long, 160-arch brick viaduct, ending high above the street and ensuring that passengers faced a hard struggle up and down many steps. He laid his rails to a gauge of 5ft, the whole track having then to be reduced to the standard gauge, creating an additional debt burden for the company. Under Braithwaite's limp direction the ECR ran out of money and failed to build the railway beyond Colchester, although it was intended to go to Norwich and Great Yarmouth.

Above: What the fuss was all about — some of the brick arches in Bishopsgate goods depot that are listed and were designed by John Braithwaite, engineer for the Eastern Counties Railway. *Brian Morrison*

Right: Remains of an old wagon turntable at Bishopsgate goods depot. The main line to Liverpool Street is over the wall to the left, about 20ft below this level. *Brian Morrison*

In 1874 a new station was opened at Liverpool Street, whereupon the old Bishopsgate terminus became offices for the goods yard. In 1964 the old station and the goods warehouses caught fire. Much of their space was occupied by customs-bonded goods such as whisky and tobacco and the circumstances of the fire were considered by some people at the time to be suspicious. The fire burned for two weeks and severely damaged the brickwork of the viaduct. Rainwater was now able to pour through into the arches below, and by 1994 the whole was in a ghastly state of decay: the under arches waterlogged, the space above a car park. It was through this great hulk of soot-encrusted, newly enforested, fire-damaged brick that London Underground wished to bring a new viaduct.

A Planning Inquiry was held at which English Heritage was represented. The inquiry occupied six weeks during October and November 1994. English Heritage made the following statement regarding the old viaduct: 'Nothing of importance stands in the path of the proposed railway. All we want to keep are the gates fronting on Shoreditch High Street.' Thus it was agreed that the new viaduct could go through the old one.

London Underground now began the work of restoring the old trackbed from Dalston Junction to Shoreditch High Street, including rebuilding eight bridges. In 1998 Railtrack's property division began renovating some of the 'Braithwaite' arches, clearing out accumulated rubbish and then letting the spaces to businesses — but only on six-month leases. Nineteen arches were fitted out variously as a cultural centre with shops, a gymnasium, swimming pool and an art centre, opened by HRH The Prince of Wales in July 2000. By the end of April 2002 LUL was ready to start its new viaduct through part of the Braithwaite viaduct, whereupon English Heritage suddenly reversed its earlier judgement and decided that the ugly old viaduct was as valuable as the late-lamented Euston Arch. On 12 May English Heritage persuaded Culture Secretary Tessa Jowell to issue a preservation order for the structure. While LUL believes this order covers only a section of the viaduct, English Heritage insists it appertains to the entire site.

A leaflet from English Heritage most disingenuously states that 'Despite a recent Grade II listing the viaduct now faces the threat of demolition under the guise that it stands in the way of the East London Line Extension'. It conveniently avoids telling the public that, 21 months earlier, English Heritage had agreed to the demolition. Concern is expressed for the 'vibrant businesses' under the arches. But LUL does not want to go through that section of the viaduct.

So the expensive legal battle of words goes on, and once again English Heritage has cost — and is costing — the East London Railway (and, ultimately, the British taxpayer) a lot of money. Meanwhile, in London, traffic accidents, congestion, noise and air pollution and the illnesses associated therewith continue.

With evidence of modernisation clearly visible in the background, these are the disused platforms at Bishopsgate goods depot almost 40 years after freight facilities were withdrawn. Given the value of real estate in this part of London, it is surprising that the site had not been redeveloped much earlier. *Brian Morrison*

Sheringham / Wensum Junction

In May 2001 the HSE Inspector came to Sheringham to inspect the site of the proposed restoration of the North Norfolk Railway track across a public road on the level to re-join with Railtrack's Cromer line. Afterwards the NNR General Manager took the HSE Inspector on a friendly tour of the railway. The HSE Inspector saw the nameboard 'WENSUM JUNCTION' at the gable end of the signalbox. 'That cannot be allowed to remain there,' says he; 'it is confusing, since this is Sheringham, not Wensum Junction.' The signalbox in question had once controlled Wensum Junction, Norwich, and, upon its removal to Sheringham, retained its old name boards. So, to avoid prosecution for confusing its drivers into thinking they were arriving at Norwich rather than Sheringham, the NNR had no choice but to obey the HSE's edict — a ruling from the very same Department that raised no objections to the exceedingly confusing two miles of track between Paddington and Ladbroke Grove.

The offending signalbox. On 1 April 1984 the North Norfolk Railway's 'J15' 0-6-0 (No 7564) passes the newly installed Sheringham West (Wensum Junction) 'box, which was then being prepared for operation in place of the ground frame used hitherto. *Brian Fisher*

It's the Way you Tell 'em . . .

In an age of 'spin', the good PR professional can be king.
However, when things do go wrong, they can do so spectacularly.
When the new Class 317 EMUs first encountered snow — surely
not an unexpected occurrence in Britain? — the units failed on
a regular basis. According to the railway industry insider asked
to comment, this failure was due to 'the wrong kind of snow'.
Here, in (presumably) the right kind of snow, on 8 February 1991,
one of the troubled units pauses at Huntingdon with a
southbound service to King's Cross. *Allan Mott*

Index